COMPUTER METHODS IN URBAN WATERSHED HYDRAULICS

THEODORE V. HROMADKA II,
Department of Civil Engineering,
University of California, Irvine

JAMES M. CLEMENTS,
Advanced Engineering Software,
Irvine, California

HARSH SALUJA,
Water Resources Division, Williamson and Schmid,
Irvine, California

Lighthouse Publications
• Mission Viejo, CA •

COMPUTER METHODS
IN
URBAN WATERSHED HYDRAULICS

Library of Congress catalog number 83-082787

First Lighthouse Publications edition.

ISBN- 0-914055-02-X

Printed in the United States of America
by: Best Printing Inc.

PREFACE

The Computer Aided Design Environment

With the increased availability of low cost microcomputers which provide computer execution speeds and data access capability comparable to the minicomputer, the use of Computer Aided Design Interaction or CADI in water resources engineering design is now considered the state of the art in progressive engineering. Typically, small engineering firms are forming which are competitive to larger established engineering firms in that the analysis of both small and large scale water resources projects is achieved by means of CADI with microcomputers.

This new trend in engineering has occurred due to the considerable reduction in computer costs, the introduction of CADI microcomputer software for water resources applications, and the state of the art in 'user-friendly' software techniques. Each of these factors has contributed to the availability of lower engineering costs, optimized construction costs due to the analysis of several alternate designs at a fraction of the cost of the analysis without use of computers, and reduced losses which result from design errors and miscalculations.

In this book, computer software will be presented for several hydraulic and hydrologic analysis procedures which commonly occur in the design of urban watershed flood control systems. The programs are based on the FORTRAN language which is available for most current microcomputers. CADI screens are also presented which display the sequence of prompts and the associated variable name and the assumed limits of entry values. Development of the library of CADI subroutines (see Chapter 6) needed to construct the form field displays, cursor addressing, and screen manipulations are machine dependent. Therefore the reader is referred to the appropriate computer user's guide.

The book also serves as documentation for the urban watershed hydraulics series of microcomputer software which is currently available in disc form from ADVANCED ENGINEERING SOFTWARE (AES) of Irvine, California. Consequently, the provided form fields reflect the actual program entry sequences as used in the AES software.

Organization of Book

The main objective of this book is to present FORTRAN computer programs (which can be accommodated on currently available microcomputers) for the analysis of flood control hydraulic and hydrologic problems. The emphasis of the included software is the hydraulic analysis of pressure flow and open channel flow pipeline systems, regular and irregular open channel systems, and the hydrologic analysis techniques of runoff hydrograph generation and routing through a detention basin.

Chapter 1 presents a review of the fundamentals of hydraulics used in the development of PROGRAM numbers 1 through 19. Chapter 2 presents the computer software for the development of a computer aided design capability for the design of pressure flow storm drain pipeline systems. Chapter 3 presents computer software for the analysis of open channel hydraulics in pipelines, regular channels, pipeline junctions, streetflow, and the determination of gradually varied flow profiles for channels and pipelines. Chapter 4 presents computer software for the analysis of irregular channel hydraulics and gradually varied flow profiles, and the hydrologic analysis of detention basins. Chapter 5 presents a computer program for the development of a watershed critical design storm runoff hydrograph for use in estimating peak flow rates and a time distribution of runoff volume. Chapter 6 presents a complete library of programs in subroutine form which provide a CADI environment for all of the water resources software supplied in this book.

ACKNOWLEDGEMENTS

The computer programs were developed during a research engagement at the University of California, Irvine campus, in 1976, and expanded by Advanced Engineering Software (AES), Irvine, California in 1979. The design of the user-friendly CRT text pages and the subsequent adaptation of the computer code to microcomputer systems was prepared under the direction of Jim Clements, MSCS, AES General Partner. Computer graphics and text illustrations were prepared by William V. Burchard of Williamson and Schmid, Irvine, California.

TABLE OF CONTENTS

CHAPTER 1.

FUNDAMENTALS OF FLUID FLOW

Introduction

The study of fluid mechanics and hydraulics requires an understanding of the basic fluid properties and the fundamental principles embodied in the conservation of mass, momentum, and energy. Consequently, the basic definitions and equations need to be presented prior to developing the detailed computer software which can be applied to solving engineering problems. In this chapter the necessary fundamentals of pipeline and channel hydraulics is briefly reviewed. These concepts will then be extended in subsequent chapters towards the development of comprehensive microcomputer software.

Fluid Properties

In the following paragraphs the basic fluid properties are defined. The foot-pound-seconds system is used for defining these properties.

* Mass (M)- A force of one pound will accelerate a mass of one slug at the rate of one foot/second/second (i.e., ft/sec^2).

* Weight (W)- Weight is the product of mass and gravitational acceleration:

$$\text{Weight(W) in pounds} = \text{Mass (M) in slugs} \times \text{gravitational acceleration (g) in ft/sec}^2. \tag{1.1}$$

* Specific weight (w) - Specific weight is the weight of a unit volume of the substance. Units are in lb/ft^3.

* Mass Density (p) - Mass density is the mass per unit volume.

$$p \text{ (slugs /ft}^3) = w/g \tag{1.2}$$

* Specific Gravity (sp. gr.) - Specific gravity of a substance is obtained by dividing the weight of a certain volume of the substance by the weight of an equal volume of water at 39.2^{o}F.

* Specific Volume (V)- Specific volume is the reciprocal of the mass density. It is the volume occupied by a unit mass of the substance.

* Viscosity of a fluid (μ)- Viscosity of a fluid is the property which determines the amount of its resistance to a shearing force. Viscosity is a measure of the relative ease or difficulty with which a particle of the fluid may be deformed, and is primarily due to the interaction between fluid molecules. Its units are lb-sec/ft^2.

* Kinematic Viscosity (v) - Kinematic viscosity is obtained by dividing the viscosity by the density:

$$v(ft^2/sec) = \mu/p \tag{1.3}$$

* Surface Tension - The property of liquids which is associated with capillary action in small tubes is surface tension. The surface tension of a liquid is the work per unit area which is required to bring molecules from inside the liquid to the surface to form additional surface area. (ft-lb/ft^2).

* Modulus of Elasticity - The modulus of elasticity is related to the change in pressure intensity divided by the corresponding change in volume per unit volume. Units are lbs/in^2.

* Unit Fluid Pressure (P) - Fluid pressure is transmitted with equal intensity in all directions and acts normal to any plane. Unit pressure is expressed as force divided by area. The units are in lb/ft^2.

* Pressure Head (h) - Pressure head is the height of a column of homogenous fluid that will produce a given intensity of pressure.

$$h(ft) = P(lb/ft^2)/w(lb/ft^3) \tag{1.4}$$

Fundamental Principles

* Equation of Continuity

The equation of continuity results from the principle of mass conservation. For steady flow, the mass flow rate of fluid passing all channel or pipeline sections is constant. That is, for cross section areas (A_1,A_2) and mean flow velocities (V_1,V_2)

$$p_1 A_1 V_1 = p_2 A_2 V_2 = \text{constant} \tag{1.5}$$

* The Energy Equation

The application of the thermodynamic principle of energy conservation to fluid flow results in the energy equation. The energy balance associated with steady, incompressible flow is summarized by

Energy at Section 1	+ Energy Added	− Energy Lost	− Energy Extracted	=	Energy at Section 2

$$(P_1/w + V_1^2/2g + z_1) + H_A - H_L - H_E = (P_2/w + V_2^2/2g + z_2) \quad (1.6)$$

where z_1 and z_2 are the elevations of the study points in the fluid. This simpler equation is known as the Bernoulli Equation.

* The Momentum Equation

Possibly the most difficult fluid flow equation to solve is the balance of momentum. Equating forces in the x-direction for steady, imcompressible, one-dimensional flow gives

$$F_x = M(V_2 - V_1) \quad (1.7)$$

where F_x is the summation of forces, M is the mass flow rate, and V_2 and V_1 are the uniform flow velocities exiting and entering the study control volume, respectively.

* Steady and Unsteady Uniform Flow

Steady, uniform flow is the fundamental type of flow treated in open channel hydraulics. The depth of the flow does not change during the time under consideration and flow velocities are assumed constant throughout each cross section of the channel. In unsteady, uniform flow the channel flow depth is a function of time.

* Varied Flows

Flow is varied if the depth of flow changes along the study reach of the channel. Varied flow may be further classified as rapidly or gradually varied (Fig. 1.1). If the depth of flow changes abruptly over a comparatively short distance, the flow is considered rapidly varied. A rapidly varied flow is a local phenomenon. Frequently encountered examples are hydraulic jumps and hydraulic drops.

Classification of Flow Regimes

In order to determine which of the basic principles of continuity, momentum, and energy to apply to particular problems

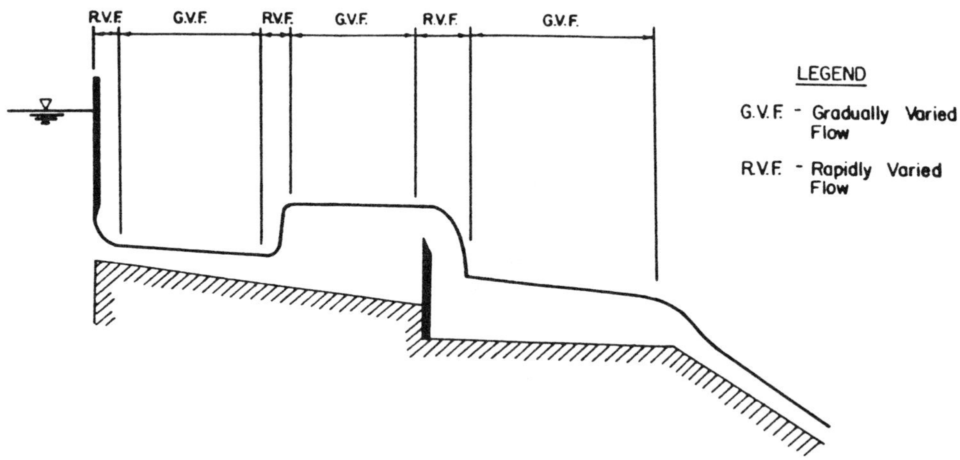

Fig. 1.1. *Combination of varied flows*

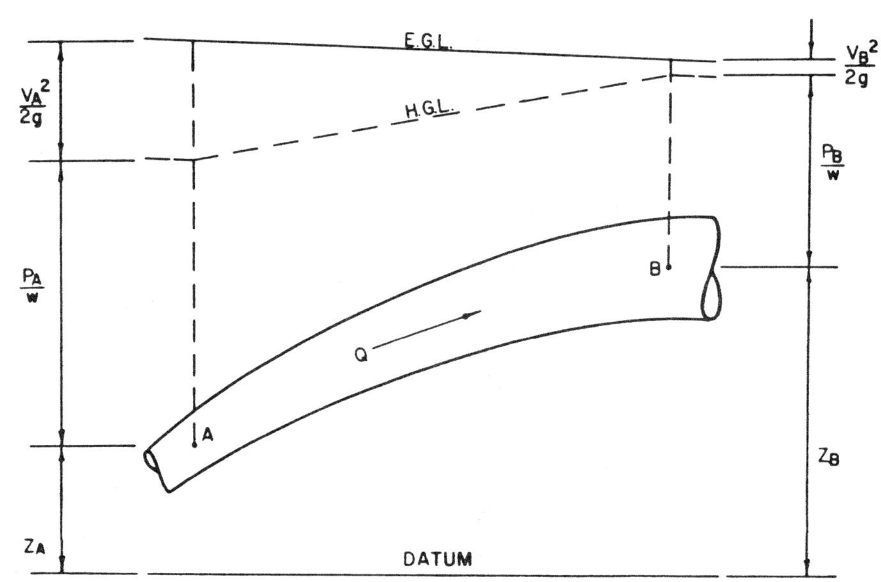

Fig. 1.2. *Pipe system energy balance*

it is necessary to classify the flow regime. For example, flow may be classified as steady or unsteady with respect to time. Other classifications of flow include rotational and irrotational, supercritical or subcritical, laminar or turbulent.

* Steady and Unsteady Flow

A steady flow is the type of flow regime where the flow velocity at any point remains constant with respect to time. Thus, although the velocity is constant with respect to time, it may vary with respect to distance.

The flow is unsteady when conditions at any point change with respect to time. Typical unsteady flow problems are the calculation of water hammer effects in pipes and translatory waves in open channels.

* Laminar and Turbulent Flow

In laminar flows, fluid particles move along smooth paths or layers. Typically, laminar flow is governed by Newton's Law of Viscosity.

In turbulent flow the fluid particles move in very irregular paths and the flow is characterized by pulsatory crosscurrent velocities. Although the effects of viscosity are still present in turbulent flow, the turbulent shear stresses are generally dominant.

* Uniform and Non-uniform Flow

The flow is uniform when all parallel cross-sections through the prismatic conduit are identical and the average velocity at each cross section is the same at any given instant. The flow is non-uniform when the velocity vector varies from place to place at any given instant.

* Rotational and Irrotational Flow

The flow is called rotational if the fluid particles within a region have a rotation about any axis. If the fluid within a region has no rotation, the flow regime is called irrotational.

* Critical, Subcritical and Supercritical Flow

The effect of gravity upon the state of flow is represented by the ratio of inertial forces to gravity forces. This ratio is given by the Froude Number, F, defined by

$$F = V / (gL)^{.5} \qquad (1.8)$$

where V is the mean velocity of flow in ft/sec., g is the acceleration of gravity in ft/sec.^{2}, and L is a characteristic length in feet.

For the special case when F is equal to 1, the flow is called critical. If F is less than or greater than 1, the flow is classified as subcritical or supercritical, respectively.

Fundamentals of Hydraulics

* Hydraulic Grade Line and Energy Grade Line

For any point in the fluid, the summation of the elevation plus the pressure head is known as the piezometric head. The piezometric head represents the level to which liquid will rise in a piezometer tube. The line drawn through the tops of a series of piezometer columns is known as the hydraulic grade line (HGL). The energy grade line (EGL) is determined by the sum of the HGL and the velocity head ($V^2/2g$) such as is shown in Fig. 1.2.

* Specific Energy

In open channel flow, the specific energy, S_E, is given by

$$S_E = y\cos^2 a + cV^2/2g \qquad (1.9)$$

where

- y = vertical depth of flow
- a = angle of the longitudinal bed profile with respect to the horizontal. (In most cases a is small, therefore $\cos^2 a = 1$)
- c = kinetic energy correction factor. This is equal to one when the velocity distribution is uniform.
- V = average flow velocity
- g = gravitational acceleration

Given the flow rate (Q), and cross section flow area (A), and for $\cos^2 a = 1$,

$$S_E = y + Q^2/2gA^2 \qquad (1.10)$$

$$(S_E - y)A^2 = Q^2/2g = \text{constant} \qquad (1.11)$$

From Equation (1.11), it is clear that the specific energy curve of Fig. 1.3 has the two asymptotes of $y = S_E$, and $y = 0$.

Alternate depths are defined as the two possible depths of flow for a given Q and S_E, and represent the two possible regimes of flow. For a point on the upper limb of the curve (Fig. 1.3), flow has a higher depth and thus a low velocity. In this case, the flow is known as subcritical. On the lower limb of the curve the flow has a lower depth and thus a high velocity. This flow is classified as supercritical. When $dS_E/dy = 0$, the flow is critical (the location of this condition is at the crest of the curve). The depth relating to critical flow is known as the critical depth, y_c.

* The Specific Force

Consider a steady, uniform, incompressible flow in an open channel between channel section A to section B, and apply Newton's second law of motion. The second law of motion states that the change of momentum per unit time in the body is equal to the resultant of all the external forces that are acting on the body. Thus for a fixed control volume,

$$Q(\beta_B V_B - \beta_A V_A) = P_A - P_B + W\sin a - F_f \qquad (1.12)$$

where

β = momentum correction factor

P_A and P_B = resultant pressures acting on section A and B, respectively

W = Equivalent weight of the fluid pressure enclosed between sections A and B

F_f = Total external forces (including friction) along the wetted boundary of the channel between section A and section B

a = angle of channel slope with respect to the horizontal

The pressure forces are calculated by

$$P_A = wA_A h_A \ , \ P_B = wA_B h_B \qquad (1.13)$$

where

w = the specific weight of the water

h = the distance to the centroid of the cross section below the water surface

If the difference of $W\sin a - F_f$ can be neglected and $\beta_1 = \beta_2 = 1$, then equation (1.12) can be simplified as

$$A_A h_A + Q^2/gA_A = A_B h_B + Q^2/gA_B \qquad (1.14)$$

Both sums of the terms in (1.14) involve identical components, and can be grouped together as the specific force, F_S. That is,

$$F_S = Ah + Q^2/gA \qquad (1.15)$$

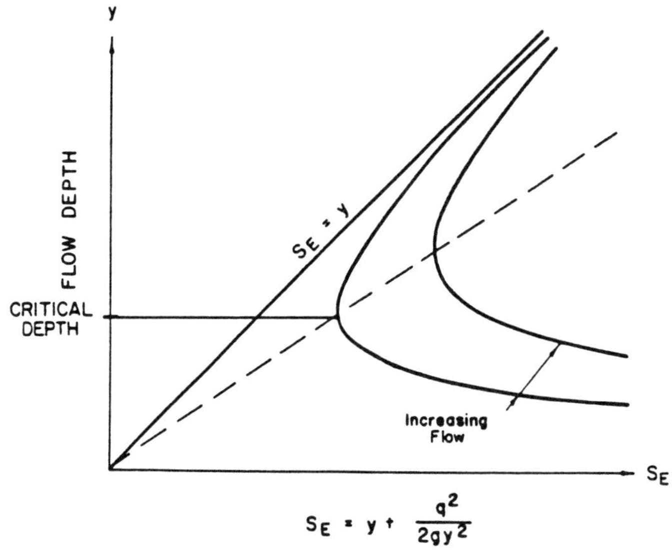

Fig. 1.3. The specific energy curve

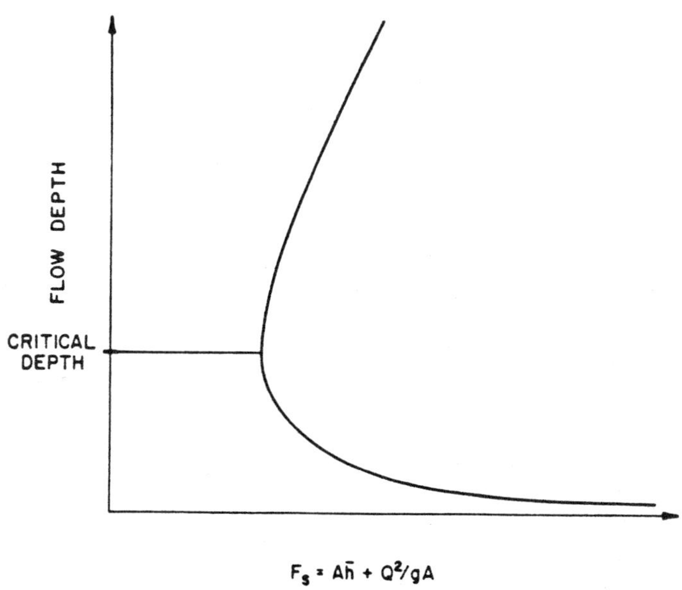

Fig. 1.4. The specific force curve

The specific force curve (Fig. 1.4) is similar in some of its characteristics to the specific energy curve (Fig. 1.3). Both the specific force and specific energy are asymptotic to the y = 0 axis. However, the specific force curve is not asymptotic to the 45° line.

* The Hydraulic Jump in a Rectangular Channel

Solution of the continuity and momentum equations for the special case of a rectangular channel leads to the following relation for the initial (y_1) and sequential depths (y_2) of a hydraulic jump on a horizontal floor:

$$y_2/y_1 = 1/2\ ((1 + 8F_1^{2})^{.5} - 1) \qquad (1.16)$$

and

$$y_1/y_2 = 1/2\ ((1 + 8F_2^{2})^{.5} - 1) \qquad (1.17)$$

Substituting these values into the energy equation gives the energy loss in the jump

$$dE = (y_2 - y_1)^3/4y_1y_2 \qquad (1.18)$$

The jump efficiency E_2/E_1 can be expressed as

$$E_2/E_1 = ((8F_1^{2} + 1)^{3/2} - 4F_1^{2} + 1)/8F_1^{2}(2 + F_1^{2}) \qquad (1.19)$$

The relative height of the jump $(y_2 - y_1)/E_1$ can be expressed as

$$(y_2 - y_1)/E_1 = ((1 + 8F_1^{2})^{.5} - 3)/(2 + F_1^{2}) \qquad (1.20)$$

The U.S. Bureau of Reclamation has classified various types of hydraulic jumps based on the Froude number. Their results are summarized below:

TABLE 1.1. HYDRAULIC JUMP CLASSIFICATIONS

F	Classification
1 to 1.7	undular jump
1.7 to 2.5	weak jump
2.5 to 4.5	oscillating jump
4.5 to 9.0	steady jump
> 9.0	strong jump

Gradually Varied Flow

Gradually varied flow can be modeled by the one-dimensional differential equation

$$dy/dx = (S_o - S_f)/(1 - F^2) \qquad (1.21)$$

where

y = flow depth
S_o = the bed slope
S_f = the friction slope
F = the Froude number

When S_f approaches S_o, dy/dx approaches zero. Therefore, water surface profiles approach the normal depth of flow asymptotically.

If F approaches unity, dy/dx approaches infinity. Therefore, by (1.21), the water surface becomes nearly vertical.

* S Profiles

A channel is classified as steep for a discharge when the normal depth is less than the critical depth, and is mild when the normal depth is greater than the critical. When the normal flow is rapid (normal depth less than critical) in a channel, the resulting profiles S_1, S_2 and S_3 are known as the steep profiles. The S_1 profile approximates gradually varied flow which is above the normal and critical depths, S_2 represents the flow profile occurring between the critical and normal depths, and S_3 occurs below the normal depth, (Fig. 1.5).

For the S_1 curve, both the numerator and denominator of (1.21) are positive and the depth increases downstream approaching a horizontal asymptote. An example is a steep canal emptying into a pool of high elevation.

For the S_2 curve, the numerator of (1.21) is negative and the denominator is positive (but approaches zero at $y = y_o$). This curve approaches the normal depth asymptotically. An example is the profile formed on the downstream side of an enlargement of a channel section.

In the S_3 curve, both the numerator and denominator of (1.21) are negative. An example is the water surface profile as the slope changes from a steep to a milder (but steep) slope.

* M Profiles

A mild slope is one where the normal flow is tranquil (i.e., normal depth, y_0, is greater than the critical depth, y_c). Three profiles may occur, and are classified as M_1, M_2, and M_3, for flow depths above normal depth, below normal and above critical depths, and below critical depth, respectively, (Fig. 1.6).

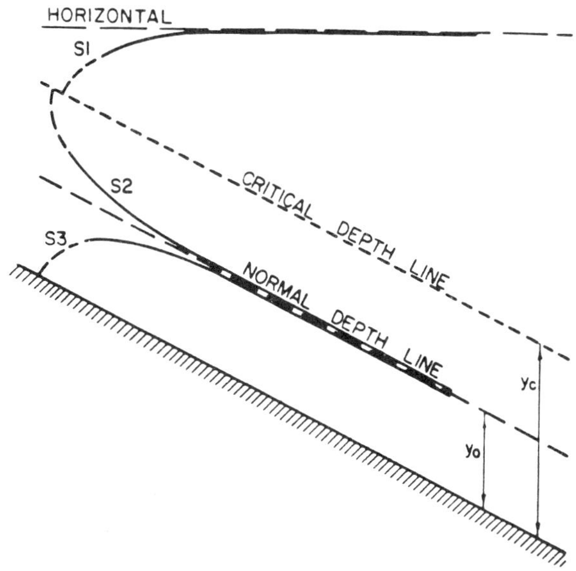

Fig. 1.5. Gradually varied flow profiles for steep slopes

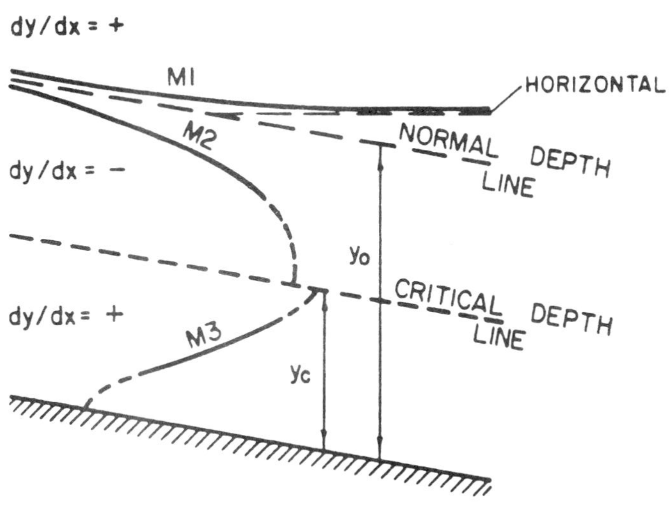

Fig. 1.6. Gradually varied flow profiles for mild slopes

For the M_1 profile $(y>y_0>y_c)$, the upstream end of the flow profile is tangent to the normal-depth line, since $dy/dx = 0$ as $y = y_0$. The downstream end is tangent to the horizontal because $dy/dx = S_0$ as y approaches infinity. A typical example in this case is the profile behind a dam in a mildly flowing river.

For the M_2 profile $(y_0>y>y_c)$, the upstream end of the flow profile is tangent to the normal depth line, since $dy/dx = 0$ as $y = y_0$. The downstream end of the flow profile is less than the normal depth but above (or equal to) the critical depth. A typical example of this profile occurs at the upstream side of a sudden enlargement of a mild channel cross-section.

For the M_3 profile $(y<y_c<y_0)$, the upstream flow depth is modeled to begin as an acute angle. The downstream flow terminates with a hydraulic jump. The most upstream flow depth is modeled as $y = 0$, and has an associated infinite flow velocity. The typical example of this profile is when a supercritical flow enters a mild channel.

* C Profiles

When the normal depth and the critical depth are equal, the profiles resulting from this are labeled C_1 and C_3. C_1 occurs when the flow depth is above the critical depth, and C_3 occurs when the flow depth is below the critical depth. These profiles represent the transition conditions between M (mild) and S (steep) flow profiles. The C_2 profile is usually associated to the case of uniform critical flow, (Fig. 1.7).

* Air Entrainment in Rapid Flow

Under high velocity flow conditions, the breaking of a large number of small waves and the intermittent action of eddies and vortices mixes air with the water. The air bubbles essentially diffuse in the flow stream by the mixing process of turbulence. In contrast, buoyancy forces these bubbles to rise to the surface. Eventual equilibrium is reached when the rates of air entrainment and release and the rates of diffusion and rise are respectively equal.

The American Society of Civil Engineers (ASCE) task committee for study of air entrainment propose a data fit relationship by the following equation

$$c = 0.743\log_{10}(S/q^{0.2}) + 0.876 \qquad (1.22)$$

where

c = the mean air concentration

S = sine of the slope angle of the flume used for the experimental study

q = the discharge per unit width in cfs/ft

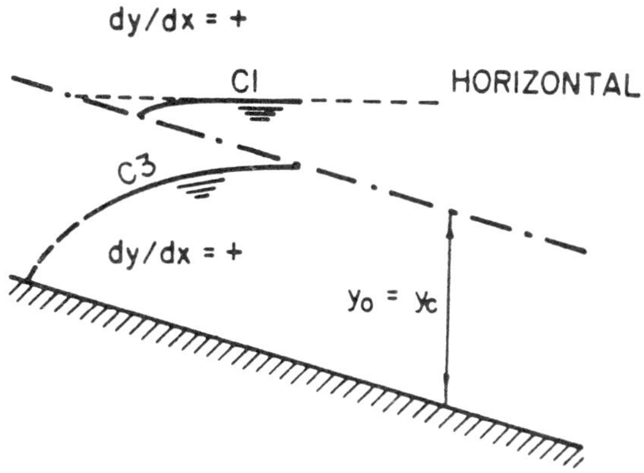

Fig. 1.7. *Gradually varied flow profiles for critical slopes*

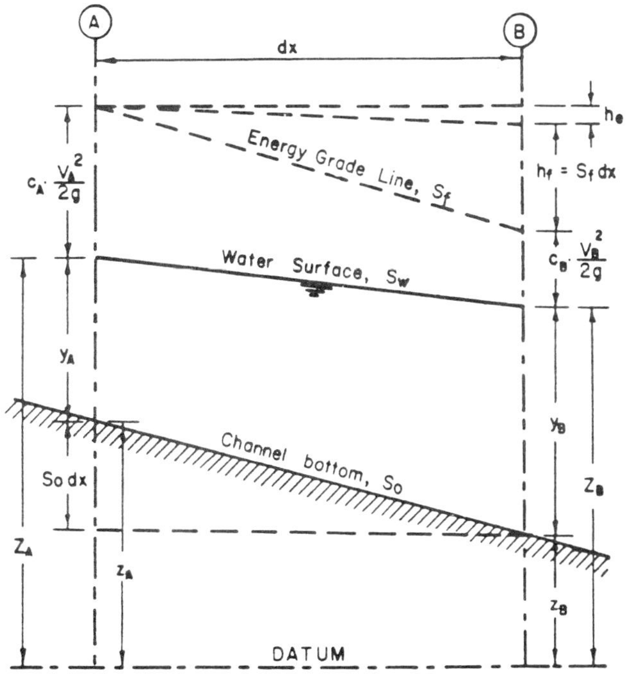

Fig. 1.8. *Channel reach used for derivation of standard step method*

A detailed discussion of this analysis is found in the ASCE task committee's report of 1961. The Corps of Engineers has modified (1.22) by

$$c = 0.701\log_{10}(S/q^{0.2}) + 0.971 \qquad (1.23)$$

* The Standard Step Method

Gradually varied flow profiles are generally computed by using any of three popular methods, namely, the graphical-integration method, the direct-integration method, and the standard step method. The standard step method continues to be the most commonly used.

In the standard step method, the computation of the flow depth is carried out on a station to station basis where the hydraulic characteristics are known. The computation procedure is a trial and error method to balance the energy equation.

For convenience, the position of the water surface is measured with respect to a horizontal datum. The water surface elevations above the datum at the two end sections can be expressed as (Fig. 1.8)

$$Z_A = y_A + z_A \qquad (1.24)$$

and

$$Z_B = y_B + z_B \qquad (1.25)$$

The friction losses are estimated between points A and B by

$$h_f = S_f\, dx = (S_A + S_B)dx/2 \qquad (1.26)$$

where S_f can be taken as the average of the friction slopes at the two end sections. The total head at sections A and B can be equated by the energy equation

$$S_0 dx + y_A + c_A V_A^2/2g = y_B + c_B V_B^2/2g + S_f dx + h_e \qquad (1.27)$$

By substitution, the following is written

$$Z_A + c_A V_A^2/2g = Z_B + c_B V_B^2/2g + h_f + h_e \qquad (1.28)$$

where h_e is the eddy loss defined by

$$h_e = k(cV^2/2g)$$

where

k = 0 to 0.1 for gradually converging reaches
k = 0 to 0.2 for gradually diverging reaches
k = 0.5 for abrupt expansion and contraction
k = 0 for prismatic and regular channel

The total heads at the two end sections A and B are

$$H_A = Z_A + c_A V_A^2/2g \tag{1.29a}$$

and

$$H_B = Z_B + c_B V_B^2/2g \tag{1.29b}$$

Using (1.29a,b), equation (1.28) can be expressed as

$$H_A = H_B + h_f + h_e \tag{1.30}$$

Given the values of H_A (or H_B), the energy head for H_B (or H_A) is computed by estimating possible flowdepths until the governing energy equations of (1.30) and (1.6) are satisfied.

Chapter Bibliography

U. S. Corps of Engineers, "Hydraulic Design of Flood Control Channels", Engineering Manual No. EM 1110-02-1601, July, 19770.

CHAPTER 2.

STORM DRAIN PRESSURE FLOW CADI MODEL

Introduction

A common approach for the development of an urban watershed flood control system is the design of a pipeflow storm drain network. A strategy for the design of the storm drain and the selection of pipe sizes is to attempt to achieve pressure flow conditions whenever possible. This strategy is especially useful in regions where the land topography has a mild gradient (approximately 0.0010 to 0.0040 ft/ft). The resulting system would be somewhat optimized in that the design consideration of the hydraulic grade line or HGL would closely conform to the maximum allowable value while providing a reasonable design for flood control protection purposes.

In this chapter, computer code for the development of a CADI software system is presented in subroutine form. Each subroutine represents a separate pressure pipeflow analysis procedure which computes the change in the energy grade line or EGL due to the specific flow process or processes being considered. For example, the friction loss subroutine computes an estimate of the change in the EGL as a function of the user entered data of pipe length, friction factor, and upstream and downstream elevations. Storing the EGL and HGL which is identified at the downstream point of the study reach, the HGL and EGL can be estimated at the upstream point of the study reach by adding the friction loss to the downstream EGL value. As in all the model processes, the HGL for the upstream point of the study reach is computed by subtracting the pressure flow condition velocity head Hv (where $Hv = V^2/2g$, and $V = Q/A$, where A is the pipe cross section area) from the EGL, i.e. HGL = EGL - Hv. In this fashion, the engineer can proceed in the upstream direction along the pipeline, calling the appropriate subroutine analysis procedure to adjust the HGL and EGL as the study progresses.

The CADI approach would involve linking all the subroutines together by a Main Menu program which allows the engineer to branch to the desired subroutine analysis procedures. Because the entire hydraulic analysis direction is in the upstream direction, the total memory of the various analysis steps is summarized by the position values of the HGL and the EGL. Consequently, the computer program operates only upon these two quantities as the study proceeds in the upstream direction. At each point of the storm drain, the engineer decides whether the computed data is acceptable for design purposes. If acceptable, the engineer ACCEPTS the most recent data values entered during the current subroutine data prompt sequence and the program automatically stores the data and returns to the Main Menu. If the engineer REJECTS the computed results, the program automatically erases the recently entered data values and returns to the Main Menu. Depending on the engineer's decision, the HGL and EGL are adjusted accordingly. This interface between the software and the engineer allows for the efficient design of the storm drain system on the first pass of data entry. The computer program does not replace the engineering design process, but amplifies the engineer's capability by removing the tedious calculation phase of the study while incorporating the design capacity of the engineer.

After the analysis of the main line storm drain is completed, the connecting storm drain laterals which merge with the main line drain at junction points can also be analyzed as separate problems. Using the main line HGL as the hydraulic control for the lateral drain, an analysis is developed in the upstream direction along the lateral analogous to the study process along the main line.

Because the CADI software determines changes in the EGL due to pressure pipeflow energy losses, the procedures are inadequate for non-pressure flow conditions. However, at locations where the storm drain unseals (i.e., the flow transitions from pressure flow conditions to non-pressure flow conditions), the program assumes soffit control (where the HGL is specified to occur at the soffit of the pipe), and the EGL is recomputed according to the new HGL and for a pressure flow condition. The engineer is informed as to this HGL adjustment, and the assumed change in the HGL is also displayed. Generally, in this condition the assumed pipesize may have been too large, in which case the engineer may reject the last computed results and attempt another design approach.

The STORM DRAIN PRESSURE FLOW CADI MODEL is composed of a Main Menu program and nine subroutine analysis procedures. The model is developed by linking the Main Menu selection program to each subroutine in order to enable the engineer to branch to the desired analysis procedure when optioned. The various programs are listed in Table 2.1.

TABLE 2.1. STORM DRAIN PRESSURE FLOW CADI MODEL PROGRAMS

PROGRAM	Description	Name
1	Main Menu	MAIN
2	friction losses	HFA
3	manhole losses	HMA
4	bend losses	HBA
5	sudden enlargement losses	HEA
6	junction losses	HJA
7	angle point losses	HAPA
8	sudden contraction losses	HCA
9	catch basin losses	HCBA
10	transition losses	TRANSA

In the following, descriptions for each subroutine used in the CADI model are presented which also include the relationships used to compute the various energy losses. These relationships were obtained from design manuals which are referenced in the Chapter Bibliography. Also included in the discussions are general design considerations which may be useful in the development of a pressure flow storm drain system. It is noted, however, that design criteria varies between agencies and that the programmed energy loss computations may be inappropriate. Generally, the necessary changes to the programs required to accommodate the local design criteria can be accommodated by simply changing the various program coefficients used to compute the energy losses. The engineer should review the local agency design criteria in order to verify the basic relationships used in the supplied programs.

PROGRAM 1. Main Menu

Before initiating the CADI program analysis, the engineer must prepare the necessary information about the storm drain system in order to enter the data values prompted by the program. The data preparation process is simply the procedure of assigning nodal points along the storm drain system. Although the actual node numbers have no impact upon the program, it is useful to number the nodes sequentially along the pipeline in a single direction. Should additional node numbers be needed at a later time, decimal points can be added to the neighboring integer node numbers (for example, 2.1, 2.2, and so forth). The storm drain system should provide the information of estimated pipeline flowline elevations, estimated pipe sizes (which may be changed during the study process), friction factors (Manning's equation), ground surface elevations at critical points or node numbers, pipeflow quantities (cubic feet per second or cfs), lengths of pipe reaches, bend and angle point angles, and the identification of a suitable hydraulic control at the most downstream point of the system (otherwise soffit control will be assumed by the program).

After the data preparation phase is completed, the CADI study phase can be initiated. The first data entry process is the definition of the hydraulic control HGL elevation at the most downstream node number of the system. The assumed pipe diameter, pipeflow, pipe flowline elevation, and node number is requested. Based on the entered data, the program establishes the control HGL and EGL.

The program continues by advancing to the Main Menu. The engineer proceeds with the analysis in the upstream direction along the pipe system. The downstream and upstream node numbers are entered as a reference to the position of the study. The engineer then selects the analysis procedure from the provided nine subroutines which determines the necessary energy losses in the pipeline reach under study. Should several analysis procedures be required in a single reach, the appropriate subroutines are called sequentially so that the sum of the individual energy losses equals the total energy loss of the system.

After the selection of a subroutine, the program branches to that analysis procedure. Once the required subroutine data is entered, the appropriate energy loss is estimated and the upstream node HGL and EGL are displayed. Also displayed is the prompt for the engineer to ACCEPT or REJECT the recent analysis results. Should the engineer ACCEPT the results, the recently entered data is stored in a data file (for future use and editing). Otherwise, the REJECT option results in the elimination of the most recent analysis. In both cases, the program returns to the Main Menu for the selection of the next analysis procedure.

PROGRAM 2. Friction Losses

Friction losses for pressure pipeflow conditions are computed from Manning's equation for steady flow

$$Q = 1.486 \times A \times R^{.667} \times S_f^{.5} / n \qquad (2.1)$$

where

- Q = flow rate, in cfs
- n = friction factor (see Table 2.2)
- A = cross section area of pipe, in ft^2
- R = hydraulic radius (ratio of area to pipe circumference)
- S_f = friction slope

For storm drain design purposes, the friction factor is usually assumed to be a constant (regardless of the flow rate). Typical values for the friction factor are given in Table 2.2.

TABLE 2.2. TYPICAL MANNING'S FRICTION FACTORS

Conduit Description	n
reinforced concrete pipe (RCP)	0.013
asbestos cement pipe (ACP)	0.012
corrugated metal pipe (CMP)	0.024
asphalt lined CMP	0.015

The increase in the EGL due to friction losses is given by the energy head loss, H_f, which is estimated by

$$H_f = S_f \times L \qquad (2.2)$$

where L is the length of pipe (ft) in the reach under study.

PROGRAM 2 computes the pressure pipeflow friction loss H_f by using the above equations. The engineer enters the downstream and upstream nodal point elevations, the length of pipe between the nodal points, a friction factor, and the flowrate. This data is then used to compute the EGL at the upstream node by

$$\text{EGL(upstream node)} = \text{EGL(downstream node)} + H_f \qquad (2.3)$$

PROGRAM 3. Manhole Losses

Manhole structures are generally constructed along the storm drain line in order to provide an adequate maintenance access to the pipeline. The loss in energy due to the passage of flow through the manhole, Hm, can be estimated as a function of the flow velocity head by

$$Hm = Km \times Hv \qquad (2.4)$$

where Km is a coefficient which is experimentally determined. The computer program computes Hm by assuming Km is the constant value, Km = 0.05. The EGL at the upstream node is computed as

$$\text{EGL(upstream node)} = \text{EGL(downstream node)} + Hm \qquad (2.5)$$

The spacing of manhole locations is generally a function of pipe size. For example, for pipe diameters less than 30 inches, manholes may be spaced at 200 to 300 foot intervals. Spacings of about 400 feet may be suitable for pipe sizes between a 30 and 45 inch diameter. And for pipe sizes larger than a 45 inch diameter, a spacing of 500 feet may be adequate.

PROGRAM 4. Bend Losses

Similar to the energy losses due to flow through manholes, the energy losses due to flow through pipeline bends, Hb, is assumed to be a function of the flow velocity head by

$$Hb = Kb \times Hv \qquad (2.6)$$

where the coefficient Kb is experimentally determined. In PROGRAM 4, the value of Kb is assumed to be independent of the flowrate and a function of the bend angle by

$$Kb = 0.25(angle/90)^{.5} \qquad (2.7)$$

where 'angle' is the central angle (less than 90 degrees) of the bend, (see Fig. 2.1).

Because the analysis proceeds from node to node, PROGRAM 4 includes the effects of both the bend and friction in the EGL computation for the upstream node. Thus, the EGL at the upstream node is estimated by

$$EGL(upstream\ node) = EGL(downstream\ node) + H_f + Hb \qquad (2.8)$$

PROGRAM 5. Sudden Enlargement Losses

As the CADI study proceeds in the upstream direction, the reduction in pipesize (in the upstream direction) without a transition structure is equivalent to a 'sudden enlargement' of pipesize in the downstream direction. The energy head loss due to a sudden enlargement of pipesize, He, is given by the relationship

$$He = (V1-V2)^2/2g \qquad (2.9)$$

where V1 is the upstream flow velocity and V2 is the downstream flow velocity. Due to the sudden enlargement occurring for the flow direction proceeding downstream, V1 is greater than V2. The EGL for the upstream node is computed as

$$EGL(upstream\ node) = EGL(downstream\ node) + He \qquad (2.10)$$

It must be noted that although the study is proceeding in the upstream direction, the sudden enlargement analysis procedure accounts for the increase in pipesize occurring in the downstream direction, (see Fig. 2.2).

PROGRAM 6. Junction Losses

The most complex analysis procedure included in the STORM DRAIN PRESSURE FLOW CADI MODEL is the pressure flow junction analysis.

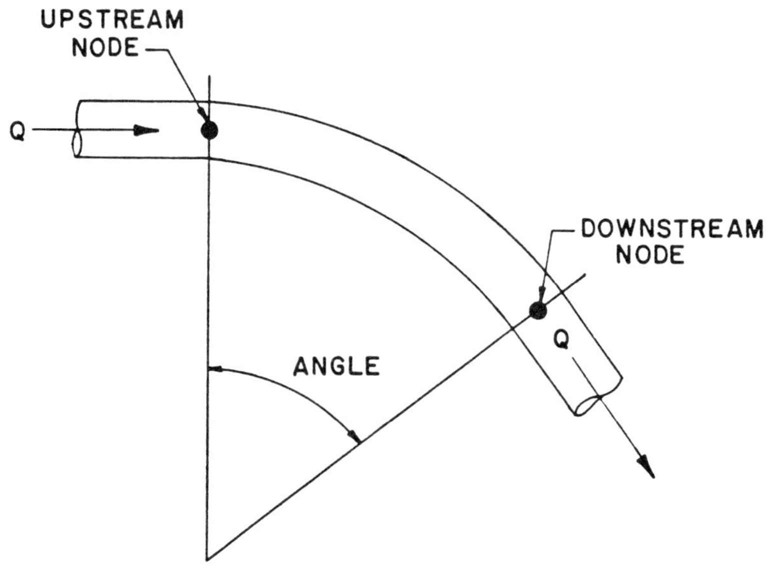

Fig. 2.1. Bend loss model geometry

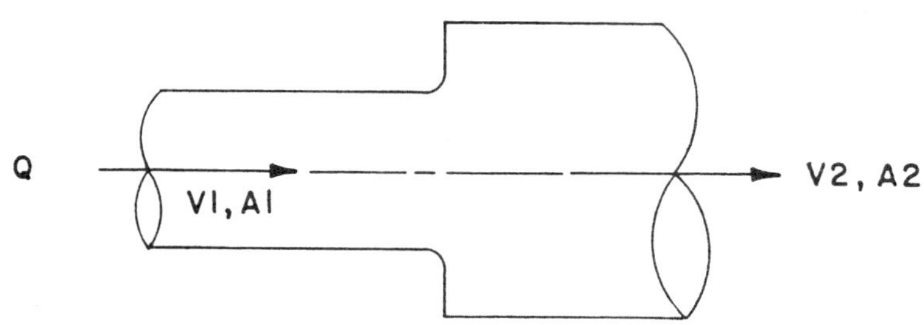

Fig. 2.2. Sudden enlargement model geometry

This program estimates the change in EGL at a junction of pressure pipeflows where up to two laterals may confluence with the mainline flow, and the junction may also serve as an inlet for flows entering from the top of the junction structure. Both the mainline and laterals may have various angles of confluence. Because the program accomodates all energy losses between nodal points, friction losses through the structure are included. Additionally, entrance losses are included should there be inlet flows.

Friction losses are computed using Eq. (2.2) to estimate the friction slope for both the upstream and downstream reaches. Using the average of the two friction slopes, the friction loss, H_f, is computed based on the length of the junction structure. Should flows enter the junction structure through an inlet constructed at the top of the structure, an additional entrance loss may be included. For example, the entrance loss may be estimated as a loss associated to a catch basin inlet, Hcb (see PROGRAM 9), where

$$Hcb = 0.20 \times Hv \qquad (2.11)$$

The junction losses due to the merging of flows of a mainline flow with one or two lateral pipeline flows may be estimated by a pressure plus momentum analysis. For example, the City of Los Angeles' Thompson equation relates the energy loss, Hj, to the change in the HGL (with friction loss H_f) by

$$Hj = dHGL + Hv1 - Hv2 + Hcb + H_f \qquad (2.12)$$

where dHGL is the change in the HGL estimated by

$$dHGL = 2(Q2V2 - Q1V1\cos(ang1) - Q3V3\cos(ang3) - Q4V4\cos(ang4)) / g(A1 + A2) \qquad (2.13)$$

where

Q1,V1,A1 = upstream flow rate, flow velocity, and pipe area
Q2,V2,A2 = downstream flow rate, flow velocity, and pipe area
Q3,V3 = lateral flow rate, and flow velocity
Q4,V4 = lateral flow rate, and flow velocity
ang1 = angle of confluence between upstream and downstream pipes
ang3,ang4 = angles of confluence between laterals and downstream pipes

In Eq. (2.12), Hv1 and Hv2 refer to the velocity heads for the upstream and downstream pressure pipeflows, respectively, and Hcb follows from Eq. (2.11). The EGL for the upstream node is computed by summing the various losses which are assumed to occur through the junction structure,

$$EGL(upstream\ node) = EGL(downstream\ node) + Hj \qquad (2.14)$$

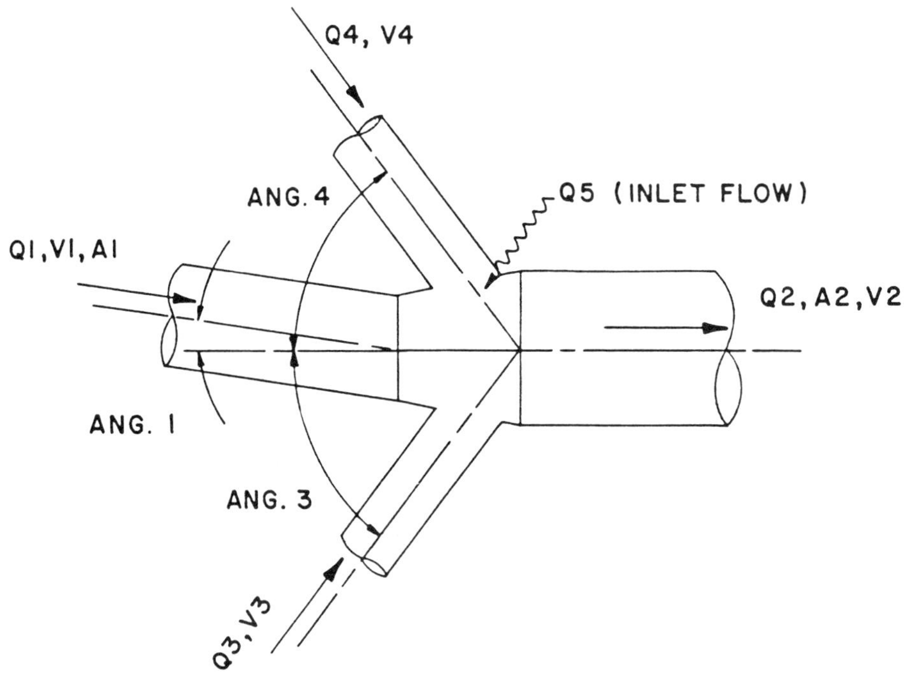

Fig. 2.3. Junction loss model geometry

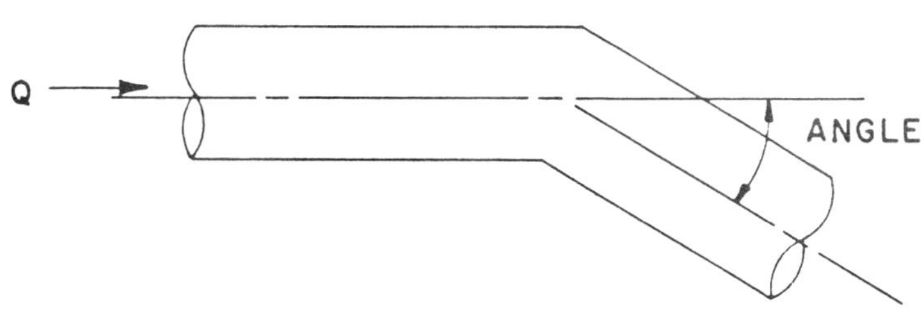

Fig. 2.4. Angle point loss model geometry

It is noted that the junction loss relations do not include losses resulting from a transition structure. This type of analysis is addressed in PROGRAM 10. Another consideration is that Eq. (2.13) may result in the estimated junction loss, Hj, being negative. To avoid this difficulty, the computer program compares Hj to the manhole loss Hm given in PROGRAM 3. Should Hm exceed Hj, then Hm is used for Hj and a warning message is printed with the computer results.

PROGRAM 7. Angle Point Losses

Energy losses due to an angle point, Hap, are computed as a function of the pressure flow velocity head by

$$\text{Hap} = \text{Kap} \times \text{Hv} \tag{2.15}$$

where Kap is a coefficient which is experimentally determined. In PROGRAM 7, the coefficient Kap is assumed to be a function of the central angle (Fig. 2.4) as shown in Table 2.3.

TABLE 2.3. TYPICAL VALUES OF Kap

Angle (degrees)	Kap	Angle (degrees)	Kap
1	.005	10	.030
2	.008	12	.037
3	.011	15	.047
4	.014	20	.067
5	.017	25	.090
6	.020	30	.115
7	.022	35	.146
8	.024	40	.184
9	.027	45	.234

The PROGRAM linearly interpolates between point values of Kap in order to provide a smooth value of the coefficient as a function of the angle point. The EGL for the upstream side of the angle point is computed as

$$\text{EGL(upstream node)} = \text{EGL(downstream node)} + \text{Hap} \tag{2.16}$$

The tabulated values are estimates of Kap only, and replacement values can be substituted by a recoding of the linear interpolation functions used in PROGRAM 7. Although the tabulated values provide for Kap coefficients for a range of angles up to 45 degrees, many design criteria manuals suggest avoiding angle points greater than 6 degrees when designing storm drain pipelines.

PROGRAM 8. Sudden Contraction Losses

Similar to PROGRAM 5 (Sudden Enlargement Losses), this program computes the energy losses due to a sudden reduction in pipesize in the flow direction. Because the CADI study proceeds in the upstream direction, PROGRAM 8 would be used when the pipe sizes suddenly increase as the study proceeds in the upstream direction. A convenient procedure for estimating the sudden contraction losses, Hc, is to assume the energy head loss to be a function of the downstream velocity head, Hv2, by (Fig. 2.5)

$$Hc = Kc \times Hv2 \qquad (2.17)$$

where Kc is a coefficient related to the ratio of upstream and downstream pressure pipe flow areas A2/A1 given by Table 2.4.

TABLE 2.4. TYPICAL VALUES OF Kc

A2/A1	Kc
.10	.46
.20	.41
.30	.36
.40	.30
.50	.24
.60	.18
.70	.12
.80	.06
.90	.02
1.00	.00

In Table 2.4, A2 is the cross section area of the downstream pipe and A1 is the cross section area of the upstream pipe. Other values for Kc may be substituted into PROGRAM 8 by revising the linear interpolation formulas used to relate Kc as a function of the pipeflow area ratio.

As with the other models, the upstream node EGL is computed by

$$EGL(\text{upstream node}) = EGL(\text{downstream node}) + Hc \qquad (2.18)$$

PROGRAM 9. Catch Basin Losses

Inlets into the storm drain system (or catch basins) are often designed in anticipation that both the HGL and EGL coincide with the ponded water surface within the inlet. Consequently, the kinetic energy of flow (or velocity head) and any losses due to the entrance of the flow into the pipeline must be accounted for

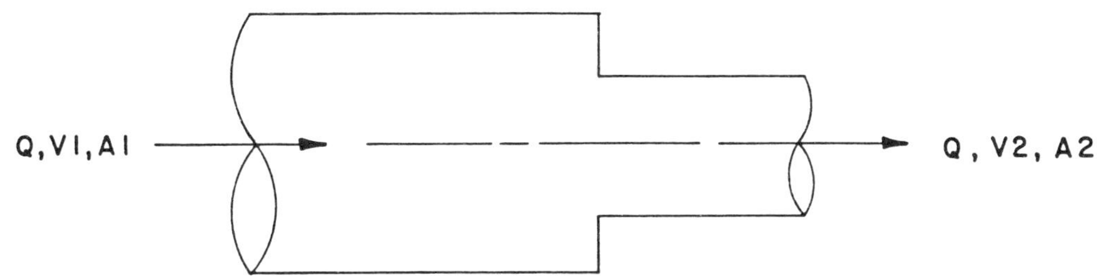

Fig. 2.5. Sudden contraction model geometry

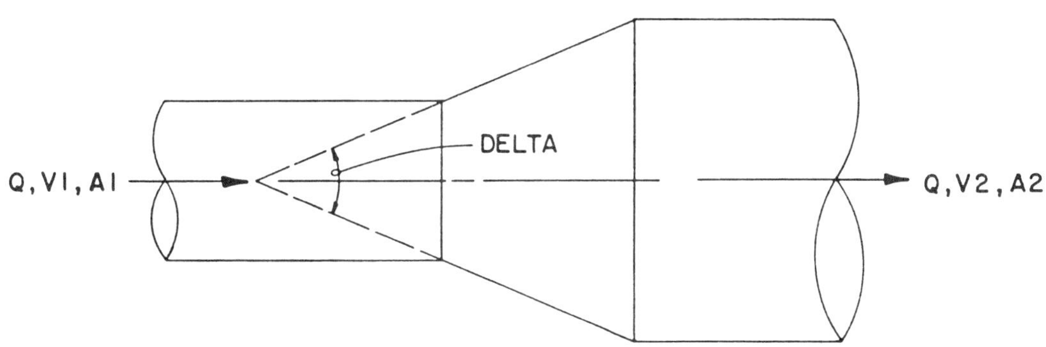

Fig. 2.6. Transition loss model geometry

in the computation of the EGL within the basin. The sum of the entrance loss and the velocity head Hv is called the catch basin loss, Hcb, and is assumed to be a function of the flow velocity head by

$$Hcb = Kcb \times Hv \tag{2.19}$$

where Kcb is a coefficient which is experimentally determined. In PROGRAM 9, Kcb is assumed to be 1.2 where 0.2 is attributed to the entrance loss. The subroutine also defines the HGL to equal the EGL in order to model a ponded water surface condition. Consequently, this subroutine may be considered to model the pipeline where the storm drain connects to the catch basin, at locations where all kinetic energy is lost due to a ponded water condition, or where angle points are greater than about 60 degrees. The upstream node EGL and HGL are computed as

$$\begin{aligned} \text{EGL(upstream node)} &= \text{EGL(downstream node)} + Hcb - Hv \quad (2.20)\\ \text{HGL(upstream node)} &= \text{EGL(upstream node)} \end{aligned}$$

PROGRAM 10. Transition Losses

Abrupt changes in pipesize are accompanied by high energy losses. In order to reduce the losses due to a sudden expansion or contraction, structures may be designed which provide for a smooth transition for the change in pipe size. The approach used in PROGRAM 10 for the estimation of the transition losses, Ht, is to model the energy losses as a function of the change in the velocity head due to the change in pipe size by

$$Ht = Kt(Hv2 - Hv1) \ldots \text{ for Hv2 greater than Hv1} \tag{2.21a}$$

$$Ht = Kt(Hv1 - Hv2) \ldots \text{ for Hv1 greater than Hv2} \tag{2.21b}$$

Equation (2.21) is assumed to apply when the transition structure wall of convergence or divergence is less than 5.75 degrees. PROGRAM 10 uses Kt = 0.1 in Eq. (2.21a), and Kt = 0.2 in Eq. (2.21b). For angles of convergence or divergence greater than 5.75 degrees, transition losses are computed by the emperical relationship

$$Ht = 3.5 \times ((\text{Tan}(\text{delta} \times .00872665))^{1.22}) \tag{2.22}$$

where delta is the total transition angle which is equal to twice the angle of convergence or divergence.

Because the transition structure also includes losses due to friction, Eq. (2.2) is used to compute the appropriate H_f value. Consequently, the EGL at the upstream node is computed by

$$\text{EGL(upstream node)} = \text{EGL(downstream node)} + Ht + H_f \tag{2.23}$$

Chapter Bibliography

"Design Manual: Channel Hydraulics and Structures," Orange County Flood Control District, OCEMA, Ca., July, 1972.

"Design Manual: Hydraulic," Los Angeles County Flood Control District, Los Angeles, Ca., 1970.

"Design Manual," Los Angeles County Road Department, Los Angeles, 1972.

PROGRAM 1: DATA ENTRY

```
Enter node number where pressure pipe flow hydraulic
control is specified................................. ===>    "ZN1"
:ALLOWABLE VALUES ARE [0.00   ] TO [9999.99   ]

Enter pipe flowline elevation of nodal point........ ===>    "ELE"
:ALLOWABLE VALUES ARE [-99999.99 ] TO [+99999.99 ]

Enter pressure flow pipe diameter(INCHES)........... ===>    "D"
:ALLOWABLE VALUES ARE [3] TO [240    ]

Enter pressure pipe flow(CFS)....................... ===>    "Q"
:ALLOWABLE VALUES ARE [0] TO [100000]

Enter assumed hydraulic grade line(HGL) at nodal
point .............................................. ===>    "HGL"
(NOTE: EGL IS ENERGY GRADE LINE)
:ALLOWABLE VALUES ARE [-99999.99 ] TO [+99999.99 ]



--------------------------------------------------------------------------------
TYPE: EXIT to leave program ; TOP  to go to top of page
```

```
Enter downstream node number........................ ===>    "ZN1"
:ALLOWABLE VALUES ARE [0.00   ] TO [9999.99   ]

Enter upstream node number.......................... ===>    "ZN2"
:ALLOWABLE VALUES ARE [0.00   ] TO [9999.99   ]

PRESSURE PIPE FLOW PROCESSES:
    1= Friction Losses
    2= Manhole Losses
    3= Pipe-bend Losses
    4= Sudden Pipe-enlargement
    5= Junction Losses
    6= Angle-point Losses
    7= Sudden Pipe Reduction
    8= Catch Basin Entrance Losses
    9= Transition Losses
                   Select pressure pipe flow process ===>    "KODE"

Enter pipe flowline elevation of node number........ ===>    "ELE"
:ALLOWABLE VALUES ARE [-99999.998] TO [+99999.998]

--------------------------------------------------------------------------------
TYPE: EXIT to leave program ; TOP  to go to top of page
      MAIN to go to main menu
```

PROGRAM 1

```
C  ---------------------------------------------------------------------
        PROGRAM PRESSURE
C  ---------------------------------------------------------------------
C
C  EXECUTIVE DRIVER FOR PRESSURE BATCH SYSTEM
C
        COMMON/NUT/NUT
C
C
        NUT=6
        HGL=0.
        EGL=0.
C
C  OPEN FILES
        OPEN 5,"PRESS.DAT"
        OPEN 6,"PRESS.ANS"
C
C  PROCESS DATA FILE
C
C  PROBLEM CONTROLS FIRST
        CALL INITP(HGL,EGL,D,Q)
C
C  SEGMENTS
100     READ FREE(5,END=1000)ZN1,ZN2,KODE,ELE
        WRITE(NUT,601)
        WRITE(NUT,600)ZN1,ZN2,KODE,ELE
        WRITE(NUT,601)
C  SEGMENTS
        IERR=0
        IF(KODE.EQ.1)CALL HFA(TEST,HGL,EGL,ERROR)
        IF(KODE.EQ.2)CALL HMA(TEST,HGL,EGL,ERROR)
        IF(KODE.EQ.3)CALL HBA(TEST,HGL,EGL,ERROR)
        IF(KODE.EQ.4)CALL HEA(TEST,HGL,EGL,ERROR)
        IF(KODE.EQ.5)CALL HJA(TEST,HGL,EGL,ERROR)
        IF(KODE.EQ.6)CALL HAPA(TEST,HGL,EGL,ERROR)
        IF(KODE.EQ.7)CALL HCA(TEST,HGL,EGL,ERROR)
        IF(KODE.EQ.8)CALL HCBA(TEST,HGL,EGL,ERROR)
        IF(KODE.EQ.9)CALL TRANSA(TEST,HGL,EGL,ERROR)
        IF(IERR.NE.0)GO TO 50
        GO TO 100
C
C  ERROR PROCESSING
50      WRITE(NUT,602)
1000    CONTINUE
C  FORMATS
600     FORMAT(3X,'NODE ',F8.2,' : HGL= <',F9.3,'>;EGL= <',F9.3,'>;',
     .  'FLOWLINE= <',F9.3,'>')
601     FORMAT(1X,76('*'))
602     FORMAT(1X,'***FATAL ERROR - CHECK DATA INPUT***')
        RETURN
        END
```

```
C ---------------------------------------------------------------------
        SUBROUTINE INITP(HGL,EGL,D,Q)
C ---------------------------------------------------------------------
C
C  PROBLEM CONTROLS
C
        COMMON /NUT/NUT
C
C
C..READ DATA INPUT
        READ FREE(5)ZN1,ELE,D,HGL,Q
        ZN2=ZN1
C
C
        WRITE(NUT,6000)
        WRITE(NUT,5018)
        WRITE(NUT,6000)
C
        WRITE(NUT,5020) ZN1,ELE,D,Q
        WRITE(NUT,5001) HGL
        WRITE(NUT,7000)
        A = .7854*D*D/144.
        V=Q/A
        HV=V*V/64.4
        TEST = ELE+D/12.
        XHGL=HGL
        IF(HGL .GE. TEST) GO TO 30
        WRITE(NUT,6500)
        XHGL=TEST
30      EGL=XHGL+HV
        WRITE(NUT,1000) ZN1,XHGL,EGL,ELE
C
C  LOAD HGL WITH XHGL TO CONTINUE CASCADING LOGIC (INPUT HAS BEEN
C  PRESERVED)
        HGL=XHGL
C
C   FORMATS
C
1000    FORMAT(3X,'NODE ',F8.2,' : HGL= <',F9.3,'>;EGL= <',F9.3,'>;',
     .  'FLOWLINE= <',F9.3,'>')
5020    FORMAT(3X,'NOTE: STEADY FLOW HYDRAULIC HEAD-LOSS COMPUTATIONS',
     .  ' BASED ON THE MOST',/,3X,'CONSERVATIVE FORMULAE FROM THE ',
     .  'CURRENT LACRD,LACFCD, AND OCEMA',/,3X,'DESIGN MANUALS.',
     .  //,3X,'DOWNSTREAM PRESSURE PIPE FLOW CONTROL DATA:',/,
     .  3X,'NODE NUMBER = ',F8.2,11X,'FLOWLINE ELEVATION = ',F8.2,/,
     .  3X,'PIPE DIAMETER(INCH) = ',F6.2,5X,'PIPE FLOW(CFS) = ',F10.2)
5001    FORMAT(3X,'ASSUMED DOWNSTREAM CONTROL HGL = ',F9.3)
5018    FORMAT(/,11X,'PRESSURE PIPE-FLOW HYDRAULICS COMPUTER PROGRAM',
     .  ' PACKAGE',/,11X, '(Reference: LACFD,LACRD,& OCEMA HYDRAULICS',
     .  ' CRITERION)',/)
6500    FORMAT(3X,'SOFFIT CONTROL ASSUMED AT BEGINNING OF PIPE',
     .  ' SYSTEM')
7000    FORMAT(4('==================='))
6000    FORMAT(4('*******************'))
C
        RETURN
        END
```

PROGRAM 2: DATA ENTRY

```
---DATA ENTRY FOR FRICTION LOSSES---

Enter pressure pipe flow(CFS)........................ ===>     "Q"
:ALLOWABLE VALUES ARE [0] TO [1000000   ]

Enter pipe diameter(INCHES).......................... ===>     "D"
:ALLOWABLE VALUES ARE [3] TO [240   ]

Enter length of pipe(FEET)........................... ===>     "XL"
:ALLOWABLE VALUES ARE [0] TO [1000000   ]

Enter mannings friction factor....................... ===>     "XN"
:ALLOWABLE VALUES ARE [.008   ] TO [.5    ]

-----------------------------------------------------------------------------
TYPE: EXIT to leave program ; TOP  to go to top of page
```

PROGRAM 3: DATA ENTRY

```
---DATA ENTRY FOR MANHOLE LOSSES---

Enter pressure pipe flow(CFS)........................ ===>     "Q"
:ALLOWABLE VALUES ARE [0] TO [1000000   ]

Enter pipe diameter(INCHES).......................... ===>     "D"
:ALLOWABLE VALUES ARE [3] TO [240   ]

-----------------------------------------------------------------------------
TYPE: EXIT to leave program ; TOP  to go to top of page
```

PROGRAM 2

```
C  ----------------------------------------------------------------
      SUBROUTINE HFA(TEST,HGL,EGL,ERROR)
C  ----------------------------------------------------------------
C
C  FRICTION LOSSES
C
      COMMON/NUT/NUT
C
C
C..READ DATA INPUT
      READ FREE (5)Q,D,XL,XN
C
C
      TEST=D/12
      DD=D/12.
      XK=35.628*DD**2.66666667
      XK=XK*.013/XN
      SF=Q/XK*Q/XK
      XHF=XL*SF
      TEGL=EGL+XHF
      A = .7854*DD*DD
      HV = Q*Q/A/A/64.4
      THGL = TEGL-HV
      WRITE(NUT,605)
      WRITE(NUT,50)Q,D,XL,XN,Q,XK,SF,XL,SF,XHF
50    FORMAT(3X,'PIPE FLOW = ',F10.2,' CFS',5X,'PIPE ',
     C 'DIAMETER = ',F6.2,' INCHES',/,3X,'PIPE LENGTH = ',
     C F10.2,' FEET',5X,'MANNINGS N = ',F7.5,/,3X,
     C 'SF=(Q/K)**2 = ((',F10.2,')/(',F10.3,'))**2 =',F10.7,/,
     C 3X,'HF=L*SF = (',F10.2,')*(',F10.7,') =',F10.3)
      ERROR=0.
C
      HGL=THGL
      EGL=TEGL
C
C   FORMATS
C
605   FORMAT(3X,'CALCULATE PRESSURE FLOW FRICTION LOSSES(LACFCD):')
      RETURN
      END
```

PROGRAM 3

```
C  -----------------------------------------------------------------
       SUBROUTINE HMA(TEST,HGL,EGL,ERROR)
C  -----------------------------------------------------------------
C
C  MANHOLE LOSSES
C
       COMMON /NUT/NUT
C
C
C...READ OFF DATA FILE
       READ FREE(5)Q,D
C
C
       TEST=D/12
       A=.7854*D*D/144.
       V=Q/A
       HV=V*V/64.4
       XHM=.05*HV
C
       WRITE(NUT,605)
       WRITE(NUT,30)Q,D,A,V,HV,HV,XHM
30     FORMAT(3X,'PIPE FLOW = ',F10.2,' CFS',5X,'PIPE ',
     C 'DIAMETER = ',F6.2,' INCHES',/,3X,'PRESSURE FLOW AREA = ',
     C F7.3,' SQUARE FEET',/,3X,'FLOW VELOCITY = ',
     C F6.2,' FEET PER SECOND',/,3X,'VELOCITY HEAD = ',F6.3,/,
     C 3X,'HMN = .05*(VELOCITY HEAD) = .05*(',F6.3,') = ',
     C F6.3)
       ERROR=0.
       TEGL=EGL+XHM
       THGL = TEGL-HV
C
       HGL=THGL
       EGL=TEGL
C
C   FORMATS
C
605    FORMAT(3X,'CALCULATE PRESSURE FLOW MANHOLE LOSSES(LACFCD):')
C
       RETURN
       END
```

PROGRAM 4: DATA ENTRY

```
---DATA ENTRY FOR PIPE-BEND LOSSES---

Enter pressure pipe flow(CFS)........................ ===>    "Q"
:ALLOWABLE VALUES ARE [0] TO [1000000   ]

Enter pipe diameter(INCHES).......................... ===>    "D"
:ALLOWABLE VALUES ARE [3] TO [240    ]

Enter pipe bend angle(DEGREES)....................... ===>    "DELTA"
:ALLOWABLE VALUES ARE [0] TO [90     ]

Enter length of pipe(FEET)........................... ===>    "XL"
:ALLOWABLE VALUES ARE [0] TO [1000000   ]

Enter mannings friction factor....................... ===>    "XN"
:ALLOWABLE VALUES ARE [.008   ] TO [.500   ]

--------------------------------------------------------------------------------
TYPE: EXIT to leave program ; TOP  to go to top of page
```

PROGRAM 5: DATA ENTRY

```
---DATA ENTRY FOR SUDDEN PIPE ENLARGEMENT---

Enter pressure pipe flow(CFS)........................ ===>    "Q"
:ALLOWABLE VALUES ARE [0] TO [1000000   ]

Enter downstream pipe diameter(INCHES).............. ===>    "D2"
:ALLOWABLE VALUES ARE [3      ] TO [240    ]

Enter upstream pipe diameter(INCHES)................ ===>    "D1"
:ALLOWABLE VALUES ARE [3      ] TO [240    ]

--------------------------------------------------------------------------------
TYPE: EXIT to leave program ; TOP  to go to top of page
```

PROGRAM 4

```
C -----------------------------------------------------------------
        SUBROUTINE HBA(TEST,HGL,EGL,ERROR)
C -----------------------------------------------------------------
C
C  PIPE-BEND LOSSES
C
        COMMON /NUT/NUT
C
C
C..READ DATA INPUT
        READ FREE (5)Q,D,DELTA,XL,XN
C
C
        TEST=D/12
        XKB=.25*(DELTA/90.)**.5
        A=.7854*D*D/144.
        V=Q/A
        HV=V*V/64.4
        XHB=XKB*HV
        XK=35.628*(D/12.)**2.66666667
        XK=XK*.013/XN
        SF=Q/XK*Q/XK
        XHF=XL*SF
C
        WRITE(NUT,605)
        WRITE(NUT,100)Q,D,DELTA,XL,XN
100     FORMAT(3X,'PIPE FLOW = ',F10.2,' CFS',5X,'PIPE DIAMETER = ',
    C   F6.2,' INCHES',/,3X,'CENTRAL ANGLE = ',F6.3,' DEGREES',/,3X,
    C   'PIPE LENGTH = ',F10.2,' FEET',5X,'MANNINGS N = ',F7.5)
        WRITE(NUT,110)A,V,HV,XKB,XKB,HV,XHB
110     FORMAT(3X,'PRESSURE FLOW AREA = ',F7.3,' SQUARE FEET',/,
    C   3X,'FLOW VELOCITY = ',F6.2,' FEET PER SECOND',/,3X,
    C   'VELOCITY HEAD = ',F6.3,5X,'BEND COEFFICIENT(KB) = ',
    C   F6.4,/,3X,'HB=KB*(VELOCITY HEAD) = (',F6.3,')*(',F6.3,
    C   ') = ',F6.3)
        WRITE(NUT,120)XK,SF,XL,SF,XHF
120     FORMAT(3X,'PIPE CONVEYANCE FACTOR = ',F10.3,
    C   5X,'FRICTION SLOPE(SF) = ',F10.7,/,
    C   3X,'FRICTION LOSSES = L*SF = (',F10.2,')*(',F10.7,') = ',F6.3)
        ERROR = 0.
        TEGL=EGL+XHF+XHB
        THGL = TEGL-HV
C
        HGL=THGL
        EGL=TEGL
C
C   FORMATS
C
605     FORMAT(3X,'CALCULATE PRESSURE FLOW PIPE-BEND LOSSES(OCEMA):')
        RETURN
        END
```

PROGRAM 5

```
C  ---------------------------------------------------------------------
        SUBROUTINE HEA(TEST,HGL,EGL,ERROR)
C  ---------------------------------------------------------------------
C
C  SUDDEN PIPE ENLARGEMENT
C
        COMMON /NUT/NUT
C
C
C..READ DATA INPUT
        READ FREE(5)Q,D2,D1
C
C
        TEST=D1/12
        A1=.7854*D1*D1/144.
        A2=.7854*D2*D2/144.
        V1=Q/A1
        V2=Q/A2
        HV1=V1*V1/64.4
        XHE=((V1-V2)**2.)/64.4
C
        WRITE(NUT,605)
        WRITE(NUT,100)Q,D2,D1
100     FORMAT(3X,'PIPE FLOW = ',F10.2,' CFS',/,
     C  3X,'DOWNSTREAM PIPE DIAMETER = ',F6.2,
     C  ' INCHES',/,3X,'UPSTREAM PIPE DIAMETER = ',F6.2,
     C  ' INCHES')
        WRITE(NUT,110)V1,V2
110     FORMAT(3X,'UPSTREAM PRESSURE FLOW VELOCITY = ',F6.2,' FEET/SEC.'
     .  ,/,3X,'DOWNSTREAM PRESSURE FLOW VELOCITY = ',F6.2,' FEET/SEC.',
     .  /,3X,'SUDDEN PIPE-FLOW ENLARGEMENT LOSSES = ')
        WRITE(NUT,120)V1,V2,XHE
120     FORMAT(3X,'((V1-V2)**2)/64.4 = ((',F7.3,'-',F7.3,')**2/64.4 = '
     .  ,F6.3)
        TEGL = EGL +XHE
        THGL = TEGL -HV1
        ERROR=0.
C
        HGL=THGL
        EGL=TEGL
C
C   FORMATS
C
605     FORMAT(3X,'CALCULATE PRESSURE FLOW SUDDEN PIPE ENLARGEMENT',
     .  ' LOSSES(LACRD):')
C
        RETURN
        END
```

PROGRAM 6: DATA ENTRY

```
---DATA ENTRY FOR JUNCTION LOSSES---PAGE 1


Enter downstream pipe flow(CFS)..................... ===>    "Q1"
:ALLOWABLE VALUES ARE [0] TO [9999.99   ]

Enter upstream pipe flow(CFS)....................... ===>    "Q2"
:ALLOWABLE VALUES ARE [0] TO [9999.99   ]

Enter first lateral pressure pipe flow(CFS)......... ===>    "Q3"
:ALLOWABLE VALUES ARE [0] TO [9999.99   ]

Enter second lateral pressure pipe flow(CFS)........ ===>    "Q4"
:ALLOWABLE VALUES ARE [0] TO [9999.99   ]

Catch basin flow(CFS) into junction structure....... ===>    "Q5"




-----------------------------------------------------------------------------
TYPE: EXIT to leave program ; TOP  to go to top of page
```

```
---DATA ENTRY FOR JUNCTION LOSSES---PAGE 2


Enter downstream pipe diameter(INCHES).............. ===>    "D1"
:ALLOWABLE VALUES ARE [3] TO [240   ]

Enter upstream pipe diameter(INCHES)................ ===>    "D2"
:ALLOWABLE VALUES ARE [3] TO [240   ]

Enter first lateral pipe diameter(INCHES)........... ===>    "D3"
(NOTE: IF LATERAL ABSENT, ENTER 0)
:ALLOWABLE VALUES ARE [0] TO [240   ]

Enter second lateral pipe diameter(INCHES).......... ===>    "D4"
(NOTE: IF LATERAL ABSENT, ENTER 0)
:ALLOWABLE VALUES ARE [0] TO [240   ]




-----------------------------------------------------------------------------
TYPE: EXIT to leave program ; TOP  to go to top of page
                            ; BACK to go back one page
```

```
---DATA ENTRY FOR JUNCTION LOSSES---PAGE 3

Enter upstream pipe angle with respect to
 downstream pipe(DEGREES)........................... ===>    "DELTA1"
:ALLOWABLE VALUES ARE [0] TO [90    ]

Enter first lateral pipe angle with respect to
 downstream pipe(DEGREES)........................... ===>    "DELTA3"
:ALLOWABLE VALUES ARE [0] TO [90    ]

Enter second lateral pipe angle with respect to
 downstream pipe(DEGREES)........................... ===>    "DELTA4"
:ALLOWABLE VALUES ARE [0] TO [90    ]

Enter junction structure length(FEET)............... ===>    "XL"
:ALLOWABLE VALUES ARE [1] TO [100   ]



--------------------------------------------------------------------------------
TYPE: EXIT to leave program ; TOP  to go to top of page
                            ; BACK to go back one page
```

```
---DATA ENTRY FOR JUNCTION LOSSES---PAGE 4

Enter downstream pipe mannings friction factor...... ===>    "XN2"
:ALLOWABLE VALUES ARE [.008  ] TO [.500  ]

Enter upstream pipe mannings friction factor........ ===>    "XN1"
:ALLOWABLE VALUES ARE [.008  ] TO [.500  ]



--------------------------------------------------------------------------------
TYPE: EXIT to leave program ; TOP  to go to top of page
                            ; BACK to go back one page
```

PROGRAM 6

```
C  -------------------------------------------------------------
         SUBROUTINE HJA(TEST,HGL,EGL,ERROR)
C  -------------------------------------------------------------
C
C  JUNCTION LOSSES
C
         COMMON /NUT/NUT
C
C
C..READ DATA INPUT
         READ FREE(5)Q1,Q2,Q3,Q4,Q5,D1,D2,D3,D4,DELTA1,DELTA3,
      .  DELTA4,XL,XN1,XN2
C
C
         D1=D1/12.
         D2=D2/12.
         D3=D3/12.
         D4=D4/12.
         PI=.7854
         A1=PI*D1*D1
         A2=PI*D2*D2
         A3=PI*D3*D3
         A4=PI*D4*D4
         V1=Q1/A1
         V2=Q2/A2
         V3=0.
         V4=0.
         IF(A3.NE.0.)V3=Q3/A3
         IF(A4.NE.0.)V4=Q4/A4
         HV2=V2*V2/64.4
         HV1=V1*V1/64.4
         X3=COS(DELTA3*PI/45.)
         X4=COS(DELTA4*PI/45.)
         X5=COS(DELTA1*PI/45.)
         TLOSS=0.
         DELY=(Q2*V2-Q1*V1*X5-Q3*V3*X3-Q4*V4*X4)/16.1/(A1+A2)
         HE=.2*HV2
         HM=.05*HV2
         SF1=(Q1/(35.628*.013/XN1*D1**2.666667))**2.
         SF2=(Q2/(35.628*.013/XN2*D2**2.666667))**2.
         SFM=.5*(SF1+SF2)
         HF=XL*SFM
C
         WRITE(NUT,605)
         WRITE(NUT,200)Q1,D11,A1,V1,DELTA1,HV1,Q2,D21,A2,V2,HV2,Q3,
      C  D31,A3,V3,DELTA3,Q4,D41,A4,V4,DELTA4,Q5
200      FORMAT(4X,'NO. DISCHARGE  DIAMETER    AREA    VELOCITY',
      C  '   DELTA      HV'
      C  ,/,5X,'1',3X,F7.1,3X,F6.2,4X,F7.3,3X,F6.3,4X,F6.3,
      C  4X,F6.3,
      C  /,5X,'2',3X,F7.1,3X,F6.2,4X,F7.3,3X,F6.3,6X,'--',6X,F6.3,/,5X,
      C  '3',3X,F7.1,3X,F6.2,4X,F7.3,3X,F6.3,4X,F6.3,7X,'-',/,5X,
      C  '4',3X,F7.1,3X,F6.2,4X,F7.3,3X,F6.3,4X,F6.3,7X,'-',/,5X,
      C  '5',3X,F7.1,'===Q5 EQUALS BASIN INPUT===')
         WRITE(NUT,210)
210      FORMAT(/,3X,'LACFCD AND OCEMA PRESSURE FLOW JUNCTION FORMULAE',
      C  ' USED:')
         WRITE(NUT,211)
211      FORMAT(3X,'DY=(Q2*V2-Q1*V1*COS(DELTA1)-Q3*V3*COS(DELTA3)-',
```

```
     C  /,7X,'Q4*V4*COS(DELTA4))/((A1+A2)*16.1)',/)
C
        WRITE(NUT,220)XN1,XN2,SF1,SF2,SFM,XL,HF
220     FORMAT(3X,'UPSTREAM MANNINGS N = ',F7.5,/,3X,
     .  'DOWNSTREAM MANNINGS N = ',F7.5,/,
     .  3X,'UPSTREAM FRICTION SLOPE = ',F7.5,/,
     C  3X,'DOWNSTREAM FRICTION SLOPE = ',F7.5,/,
     C  3X,'AVERAGED FRICTION SLOPE IN JUNCTION ASSUMED AS ',F7.5,/,
     C  3X,'JUNCTION LENGTH(FEET) = ',F6.2,5X,'FRICTION LOSS = ',F6.3)
        IF(Q5.EQ.0.)HE=0.
        WRITE(NUT,225)HE
225     FORMAT(3X,'ENTRANCE LOSSES = ',F6.3)
        TLOSS=TLOSS+HE
230     XHJ=DELY+HV1-HV2
        IF(XHJ.GE.HM)GOTO 240
        WRITE(NUT,235)XHJ,HM
235     FORMAT(3X,'MANHOLE LOSSES GREATER THAN THOMPSON MOMENTUM LOSSES'
     C  ,/,3X,'MOMENTUM LOSSES = ',F6.3,5X,'MANHOLE LOSSES = ',F6.3)
        XHJ=HM
240     TLOSS=TLOSS+HF+XHJ
        TEGL=EGL+TLOSS
        THGL=TEGL-HV1
        WRITE(NUT,250)DELY,HV1,HV2,HF,HE,TLOSS
250     FORMAT(3X,'JUNCTION LOSSES = DY+HV1-HV2+(FRICTION LOSS)+',
     C  '(ENTRANCE LOSSES)',/,3X,'JUNCTION LOSSES = ',
     C  F6.3,'+',F6.3,'-',F6.3,'+(',F6.3,')+(',F6.3,') = ',F6.3)
        ERROR=0.
C
        HGL=THGL
        EGL=TEGL
C
C    FORMATS
C
605     FORMAT(3X,'CALCULATE PRESSURE FLOW JUNCTION LOSSES:')
C
        RETURN
        END
```

PROGRAM 7: DATA ENTRY

```
---DATA ENTRY FOR ANGLE-POINT LOSSES---


Enter pressure pipe flow(CFS)....................... ===>    "Q"
:ALLOWABLE VALUES ARE [0] TO [1000000   ]

Enter pipe diameter(INCHES)......................... ===>    "D"
:ALLOWABLE VALUES ARE [3] TO [240   ]

Enter pressure flow angle-point angle(DEGREES)...... ===>    "DELTA"
:ALLOWABLE VALUES ARE [0] TO [45    ]






-------------------------------------------------------------------------------
    TYPE: EXIT to leave program ; TOP  to go to top of page
```

PROGRAM 8: DATA ENTRY

```
---DATA ENTRY FOR SUDDEN PIPE REDUCTION---


Enter pressure pipe flow(CFS)....................... ===>    "Q"
:ALLOWABLE VALUES ARE [0] TO [1000000   ]

Enter downstream pipe diameter(INCHES).............. ===>    "D2"
:ALLOWABLE VALUES ARE [3     ] TO [240   ]

Enter upstream pipe diameter(INCHES)................ ===>    "D1"
:ALLOWABLE VALUES ARE [3     ] TO [240   ]






-------------------------------------------------------------------------------
    TYPE: EXIT to leave program ; TOP  to go to top of page
```

PROGRAM 7

```
C ----------------------------------------------------------------------
      SUBROUTINE HAPA(TEST,HGL,EGL,ERROR)
C ----------------------------------------------------------------------
C
C  ANGLE-POINT LOSSES
C
      COMMON /NUT/NUT
C
C
C..READ DATA INPUT
      READ FREE(5)Q,D,DELTA
C
C
      TEST=D/12
      A=.7854*D*D/144.
      V=Q/A
      HV=V*V/64.4
      IF(DELTA.LE.1.)XKA=.005-(1.-DELTA)*.005
      IF(DELTA.GT.1..AND.DELTA.LE.6.)XKA=.005+(DELTA-1.)*.003
      IF(DELTA.GT.6..AND.DELTA.LE.8.)XKA=.02+(DELTA-6.)*.002
      IF(DELTA.GT.8..AND.DELTA.LE.10.)XKA=.024+(DELTA-8.)*.003
      IF(DELTA.GT.10..AND.DELTA.LE.12.)XKA=.03+(DELTA-10.)*.0035
      IF(DELTA.GT.12..AND.DELTA.LE.15.)XKA=.037+(DELTA-12.)*.00333333
      IF(DELTA.GT.15..AND.DELTA.LE.20.)XKA=.047+(DELTA-15.)*.004
      IF(DELTA.GT.20..AND.DELTA.LE.25.)XKA=.067+(DELTA-20.)*.0046
      IF(DELTA.GT.25..AND.DELTA.LE.30.)XKA=.09+(DELTA-25.)*.005
      IF(DELTA.GT.30..AND.DELTA.LE.35.)XKA=.115+(DELTA-30.)*.0062
      IF(DELTA.GT.35..AND.DELTA.LE.40.)XKA=.146+(DELTA-35.)*.0076
      IF(DELTA.GT.40..AND.DELTA.LE.45.)XKA=.184+(DELTA-40.)*.01
      IF(XKA.LE.0.)XKA=0.
      HAP=HV*XKA
C
      WRITE(NUT,605)
      WRITE(NUT,100)Q,D,DELTA
100   FORMAT(3X,'PIPE FLOW = ',F10.2,' CFS',5X,'PIPE DIAMETER = ',
     C F6.2,' INCHES',/,3X,'PIPE ANGLE POINT DELTA = ',F6.2,' DEGREES')
      WRITE(NUT,103)XKA
103   FORMAT(3X,'PRESSURE FLOW ANGLE-POINT COEFFICIENT KA = ',
     C F6.4)
      WRITE(NUT,110)V,HV,XKA,HV,HAP
110   FORMAT(3X,'PRESSURE FLOW VELOCITY = ',F6.2,' FEET/SEC.',/,3X,
     C 'VELOCITY HEAD = ',F6.3,/,3X,'HAPT=KA*(VELOCITY HEAD) = ',
     C '(',F6.4,')*(',F6.3,') = ',F6.3)
      ERROR=0.
      TEGL=EGL+HAP
      THGL = TEGL-HV
C
      HGL=THGL
      EGL=TEGL
C
C   FORMATS
C
605   FORMAT(3X,'CALCULATE PRESSURE FLOW ANGLE-POINT LOSSES(LACRD):')
C
      RETURN
      END
```

PROGRAM 8

```
C ------------------------------------------------------------------
        SUBROUTINE HCA(TEST,HGL,EGL,ERROR)
C ------------------------------------------------------------------
C
C  SUDDEN PIPE REDUCTION
C
        COMMON /NUT/NUT
C
C
C..READ DATA INPUT
        READ FREE(5)Q,D1,D2
C
C
        TEST=D1/12.
        X=D2/D1*D2/D1
        IF(X.GE..1)GOTO 50
        WRITE(WT,655)
        GO TO 3020
C
50      CONTINUE
C
        IF(X.GE..1.AND.X.LT..3)XK=.46-.5*(X-.1)
        IF(X.GE..3.AND.X.LT..8)XK=.36-.6*(X-.3)
        IF(X.GE..8.AND.X.LT..9)XK=.06-.4*(X-.8)
        IF(X.GE..9.AND.X.LE.1.)XK=.02-.2*(X-.9)
        A2=.7854*D2*D2/144.
        A1=.7854*D1*D1/144.
        V1=Q/A1
        HV1=V1*V1/64.4
        V2=Q/A2
        HV2 = V2*V2/64.4
        V=V1
        HV = HV1
        IF(HV2 .GT. HV1) HV=HV2
        XHC = XK*HV
        IF(HV2 .GT. HV1) V=V2
C
        WRITE(NUT,605)
        WRITE(NUT,100)Q,D1,D2
100     FORMAT(3X,'PIPE FLOW = ',F10.2,' CFS',/,
     C  3X,'UPSTREAM PIPE DIAMETER = ',F6.2,
     C  ' INCHES',/,3X,'DOWNSTREAM PIPE DIAMETER = ',F6.2,
     C  ' INCHES')
        WRITE(NUT,110)V,HV,XK,XK,HV,XHC
110     FORMAT(3X,'PRESSURE FLOW VELOCITY = ',F6.2,' FEET/SEC.',
     C  /,3X,'VELOCITY HEAD = ',F10.2,/,3X,
     C  'PIPE REDUCTION LOSS COEFFICIENT KC = ',F6.3,/,
     C  3X,'HC(PER LACFCD)=KC*(VELOCITY HEAD) = ',
     C  '(',F6.3,')*(',F10.2,') = ',F6.3)
        ERROR=0.
        TEGL=EGL+XHC
        THGL=TEGL-HV1
C
        HGL=THGL
        EGL=TEGL
3020    CONTINUE
C
C   FORMATS
C
```

```
605     FORMAT(3X,'CALCULATE PRESSURE FLOW SUDDEN PIPE REDUCTION'
     .  ,' LOSSES(LACRD):')
655     FORMAT(3X,'*ERROR-SUDDEN PIPE SIZE REDUCTION TOO SEVERE')
C
        RETURN
        END
```

PROGRAM 9: DATA ENTRY

```
---DATA ENTRY FOR CATCH BASIN ENTRANCE LOSSES---


Enter pressure pipe flow(CFS) from catch basin...... ===>     "Q"
:ALLOWABLE VALUES ARE [0] TO [1000000   ]

Enter pipe diameter(INCHES) from catch basin........ ===>     "D"
:ALLOWABLE VALUES ARE [3] TO [240   ]









--------------------------------------------------------------------------------
TYPE: EXIT to leave program ; TOP  to go to top of page
```

PROGRAM 9

```
C ----------------------------------------------------------------------
      SUBROUTINE HCBA(TEST,HGL,EGL,ERROR)
C ----------------------------------------------------------------------
C
C  CATCH BASIN ENTRANCE LOSSES
C
      COMMON /NUT/NUT
C
C
C..READ DATA INPUT
      READ FREE(5)Q,D
C
C
      TEST=D/12
      A=.7854*D*D/144.
      V=Q/A
      HV=V*V/64.4
      HE=.2*HV
C
      WRITE(NUT,605)
      WRITE(NUT,100)Q,D
100   FORMAT(3X,'PIPE FLOW(CFS) = ',F10.2,5X,'PIPE DIAMETER(INCH) = ',
     C F6.2)
      WRITE(NUT,105)HV,HV,HE
105   FORMAT(3X,'PRESSURE FLOW VELOCITY HEAD = ',F6.3,/,
     C 3X,'CATCH BASIN ENERGY LOSS = .2*(VELOCITY HEAD) = .2*(',
     C F6.3,') =',F6.3)
      ERROR = 0.
      TEGL=EGL+HE
      THGL=TEGL
C
      HGL=THGL
      EGL=TEGL
C
C   FORMATS
C
605   FORMAT(3X,'CALCULATE PRESSURE FLOW CATCH BASIN ENTRANCE'
     . ,' LOSSES(LACFCD):')
C
      RETURN
      END
```

PROGRAM 10: DATA ENTRY

```
---DATA ENTRY FOR TRANSITION LOSSES---PAGE 1

Enter pressure pipe flow(CFS)....................... ===>    "Q"
:ALLOWABLE VALUES ARE [0] TO [1000000   ]

Enter downstream pipe diameter(INCHES).............. ===>    "D2"
:ALLOWABLE VALUES ARE [3] TO [240   ]

Enter upstream pipe diameter(INCHES)................ ===>    "D1"
:ALLOWABLE VALUES ARE [3] TO [240   ]

Enter length of transition(FEET).................... ===>    "XL"
:ALLOWABLE VALUES ARE [0] TO [1000  ]

--------------------------------------------------------------------------------
TYPE: EXIT to leave program ; TOP  to go to top of page
```

```
---DATA ENTRY FOR TRANSITION LOSSES---PAGE 2

Enter downstream pipe mannings friction factor...... ===>    "XN2"
:ALLOWABLE VALUES ARE [.008  ] TO [.500  ]

Enter upstream pipe mannings friction factor........ ===>    "XN1"
:ALLOWABLE VALUES ARE [.008  ] TO [.500  ]

Enter total-angle-of-transition(DEGREES)............ ===>    "DELTA"
(NOTE: SEE LACFCD DESIGN MANUAL CHART No. B-11
FOR DESCRIPTION.)
:ALLOWABLE VALUES ARE [0] TO [22.666666 ]

--------------------------------------------------------------------------------
TYPE: EXIT to leave program ; TOP  to go to top of page
                            ; BACK to go back one page
```

PROGRAM 10

```
C -----------------------------------------------------------------
      SUBROUTINE TRANSA(TEST,HGL,EGL,ERROR)
C -----------------------------------------------------------------
C
C  TRANSITION LOSSES
C
      COMMON /NUT/NUT
C
C
C..READ DATA INPUT
      READ FREE(5)Q,D1,D2,XL,XN1,XN2,DELTA
C
C
      TEST=D1/12.
      A1=.7854*D1*D1/144.
      A2=.7854*D2*D2/144.
      V1=Q/A1
      V2=Q/A2
      HV1=V1*V1/64.4
      HV2=V2*V2/64.4
      XK1=35.628*.013/XN1*(D1/12.)**2.6666667
      XK2=35.628*.013/XN2*(D2/12.)**2.6666667
      SF1=(Q/XK1)**2.
      SF2=(Q/XK2)**2.
      SFM=.5*(SF1+SF2)
      HF=XL*SFM
C
      WRITE(NUT,605)
      WRITE(NUT,200)Q,D1,D2,XL,XN1,XN2,DELTA
200   FORMAT(3X,'PRESSURE PIPE FLOW(CFS) = ',F10.2,/,
     C 3X,'UPSTREAM DIAMETER(INCH) = ',F6.2,5X,
     C 'DOWNSTREAM DIAMETER(INCH) = ',F6.2,/,
     C 3X,'TRANSITION STRUCTURE LENGTH(FEET) = ',F6.2,/,
     C 3X,'UPSTREAM MANNINGS N = ',F6.4,5X,
     C 'DOWNSTREAM MANNINGS N = ',F6.4,/,3X,'TOTAL-ANGLE-OF-',
     C 'TRANSITION(DEGREES) = ',F6.3)
      WRITE(NUT,202)V1,HV1,V2,HV2
202   FORMAT(3X,'UPSTREAM VELOCITY(FPS) = ',F7.3,3X,
     C 'UPSTREAM VELOCITY HEAD = ',F6.3,/,3X,
     C 'DOWNSTREAM VELOCITY(FPS) = ',F7.3,3X,
     C 'DOWNSTREAM VELOCITY HEAD = ',F6.3)
      IF(DELTA.LE.11.5)GOTO 1000
      WRITE(NUT,201)
      IF(V1.GT.V2)WRITE(NUT,300)
      IF(V2.GT.V1)WRITE(NUT,301)
201   FORMAT(3X,'DELTA EXCEEDS 11.5 DEGREES, USE LACFCD "GIBSON"',
     C ' NOMOGRAPH FOR KE VALUE')
      XKE=3.5*((TAN(DELTA*.00872665))**1.22)
300   FORMAT(3X,'VELOCITY DECREASE IN TRANSITION',/,3X,
     C 'HT = KO*(V1-V2)*(V1-V2)/64.4')
301   FORMAT(3X,'VELOCITY INCREASE IN TRANSITION',/,3X,
     C 'HT = KI*(V2-V1)*(V2-V1)/64.4')
      IF(V2.GE.V1)XKE=XKE/2.
      XHT=(V2-V1)*(V2-V1)*XKE/64.4
      GOTO 2000
1000  XKE=.2
      IF(V2.GE.V1)XKE=.1
      IF(V2.GE.V1)WRITE(NUT,305)
      IF(V1.GT.V2)WRITE(NUT,306)
```

```
305     FORMAT(3X,'HT = .1*(V2*V2-V1*V1)/64.4')
306     FORMAT(3X,'HT = .2*(V1*V1-V2*V2)/64.4')
        XHT=(ABS(V2*V2-V1*V1))*XKE/64.4
2000    WRITE(NUT,2010)XKE,XHT
2010    FORMAT(3X,'TRANSITION LOSS COEFFICIENT = ',F6.4,/,
     C  3X,'TRANSITION LOSS = ',F6.3)
        WRITE(NUT,2020)SF1,SF2,SFM,HF
2020    FORMAT(3X,'UPSTREAM FRICTION SLOPE = ',F7.5,/,3X,
     C  'DOWNSTREAM FRICTION SLOPE = ',F7.5,/,3X,
     C  'AVERAGE TRANSITION STRUCTURE FRICTION SLOPE ASSUMED = ',F7.5,/,
     C  3X,'HF = L*SF = ',F6.3)
        WRITE(NUT,2030)XHT,HF
2030    FORMAT(3X,'TOTAL LOSSES = (HT)+(HF) = (',F6.3,')+(',F6.3,')')
        TEGL=EGL+XHT+HF
        THGL=TEGL-HV1
        ERROR=0.
C
        HGL=THGL
        EGL=TEGL
C
C   FORMATS
C
605     FORMAT(3X,'CALCULATE PRESSURE FLOW TRANSITION LOSSES(LACFCD):')
C
        RETURN
        END
```

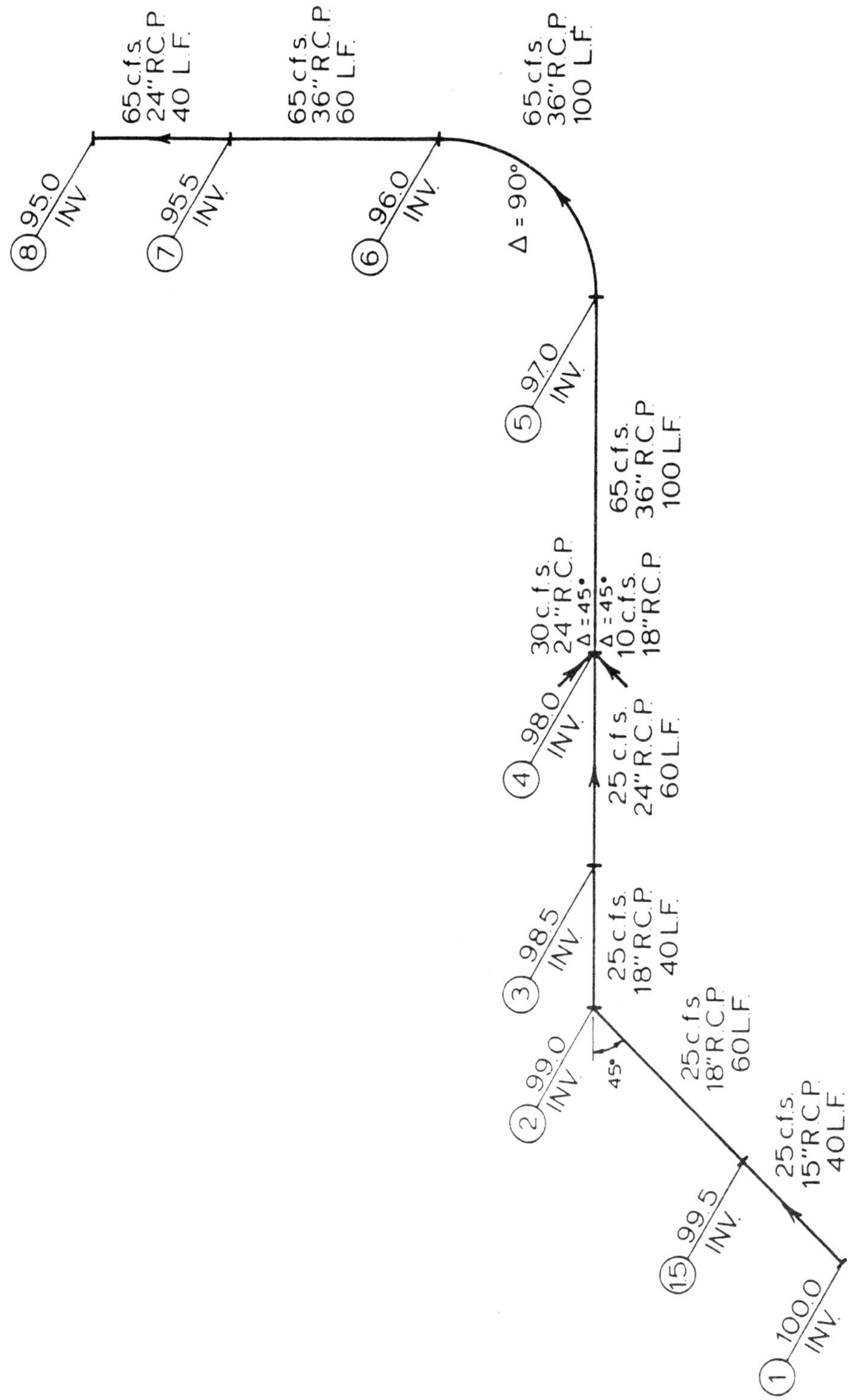

EXAMPLE PROBLEM – HYDRAULIC SCHEMATIC

```
**********DESCRIPTION OF RESULTS********************************************
*                                                                          *
*                              EXAMPLE PROBLEM                             *
*                                                                          *
****************************************************************************

****************************************************************************
   NOTE: STEADY FLOW HYDRAULIC HEAD-LOSS COMPUTATIONS BASED ON THE MOST
   CONSERVATIVE FORMULAE FROM THE CURRENT LACRD,LACFCD, AND OCEMA
   DESIGN MANUALS.

   DOWNSTREAM PRESSURE PIPE FLOW CONTROL DATA:
   NODE NUMBER =      8.00            FLOWLINE ELEVATION =    95.00
   PIPE DIAMETER(INCH) =  24.00      PIPE FLOW(CFS) =      65.00
   ASSUMED DOWNSTREAM CONTROL HGL =    97.000
============================================================================
   NODE      8.00 : HGL= <    97.000>;EGL= <  103.647>;FLOWLINE= <   95.000>


<<<<<<<<<<<<<<<<<<<<<<<<<<<<<<<<<<<<<<>>>>>>>>>>>>>>>>>>>>>>>>>>>>>>>>>>>>>>

                    Advanced Engineering Software [AES]
                              SERIAL No. B0008
                    REV. 2.0      RELEASE DATE: 7/30/82

<<<<<<<<<<<<<<<<<<<<<<<<<<<<<<<<<<<<<<>>>>>>>>>>>>>>>>>>>>>>>>>>>>>>>>>>>>>>


============================================================================
   PRESSURE FLOW PROCESS FROM NODE      8.00 TO NODE      7.00 IS CODE =  1
   UPSTREAM NODE      7.00      ELEVATION =     95.50
----------------------------------------------------------------------------
   CALCULATE PRESSURE FLOW FRICTION LOSSES(LACFCD):
   PIPE FLOW =        65.00 CFS      PIPE DIAMETER =  24.00 INCHES
   PIPE LENGTH =        40.00 FEET      MANNINGS N =  .01300
   SF=(Q/K)**2 = ((      65.00)/(    226.224))**2 =  .0825563
   HF=L*SF = (       40.00)*(  .0825563) =       3.302
   NODE      7.00 : HGL= <  100.302>;EGL= <  106.949>;FLOWLINE= <   95.500>

============================================================================
   PRESSURE FLOW PROCESS FROM NODE      7.00 TO NODE      7.00 IS CODE =  7
   UPSTREAM NODE      7.00      ELEVATION =     95.50
```

```
--------------------------------------------------------------------------
  CALCULATE PRESSURE FLOW SUDDEN PIPE REDUCTION LOSSES(LACRD):
  PIPE FLOW =       65.00 CFS
  UPSTREAM PIPE DIAMETER =  36.00 INCHES
  DOWNSTREAM PIPE DIAMETER =  24.00 INCHES
  PRESSURE FLOW VELOCITY =  20.69 FEET/SEC.
  VELOCITY HEAD =          6.65
  PIPE REDUCTION LOSS COEFFICIENT KC =    .273
  HC(PER LACFCD)=KC*(VELOCITY HEAD) = (  .273)*(        6.65) =  1.817
  NODE      7.00 : HGL= <  107.453>;EGL= <  108.766>;FLOWLINE= <   95.500>


==========================================================================
  PRESSURE FLOW PROCESS FROM NODE     7.00 TO NODE      6.00 IS CODE =  1
  UPSTREAM NODE      6.00     ELEVATION =    96.00
--------------------------------------------------------------------------
  CALCULATE PRESSURE FLOW FRICTION LOSSES(LACFCD):
  PIPE FLOW =       65.00 CFS     PIPE DIAMETER =  36.00 INCHES
  PIPE LENGTH =        60.00 FEET      MANNINGS N =  .01300
  SF=(Q/K)**2 = ((      65.00)/(    666.983))**2 =  .0094972
  HF=L*SF = (      60.00)*(  .0094972) =        .570
  NODE      6.00 : HGL= <  108.023>;EGL= <  109.336>;FLOWLINE= <   96.000>


==========================================================================
  PRESSURE FLOW PROCESS FROM NODE     6.00 TO NODE      5.00 IS CODE =  3
  UPSTREAM NODE      5.00     ELEVATION =    97.00
--------------------------------------------------------------------------
  CALCULATE PRESSURE FLOW PIPE-BEND LOSSES(OCEMA):
  PIPE FLOW =       65.00 CFS     PIPE DIAMETER =  36.00 INCHES
  CENTRAL ANGLE = 90.000 DEGREES
  PIPE LENGTH =       100.00 FEET      MANNINGS N =  .01300
  PRESSURE FLOW AREA =    7.069 SQUARE FEET
  FLOW VELOCITY =    9.20 FEET PER SECOND
  VELOCITY HEAD =  1.313     BEND COEFFICIENT(KB) =  .2500
  HB=KB*(VELOCITY HEAD) = (  .250)*( 1.313) =   .328
  PIPE CONVEYANCE FACTOR =     666.983      FRICTION SLOPE(SF) =   .0094972
  FRICTION LOSSES = L*SF = (     100.00)*(  .0094972) =   .950
  NODE      5.00 : HGL= <  109.301>;EGL= <  110.614>;FLOWLINE= <   97.000>


==========================================================================
  PRESSURE FLOW PROCESS FROM NODE     5.00 TO NODE      5.00 IS CODE =  2
  UPSTREAM NODE      5.00     ELEVATION =    97.00
--------------------------------------------------------------------------
  CALCULATE PRESSURE FLOW MANHOLE LOSSES(LACFCD):
  PIPE FLOW =       65.00 CFS     PIPE DIAMETER =  36.00 INCHES
  PRESSURE FLOW AREA =    7.069 SQUARE FEET
  FLOW VELOCITY =    9.20 FEET PER SECOND
  VELOCITY HEAD =  1.313
  HMN = .05*(VELOCITY HEAD) = .05*( 1.313) =    .066
  NODE      5.00 : HGL= <  109.367>;EGL= <  110.680>;FLOWLINE= <   97.000>


==========================================================================
  PRESSURE FLOW PROCESS FROM NODE     5.00 TO NODE      4.00 IS CODE =  1
  UPSTREAM NODE      4.00     ELEVATION =    98.00
--------------------------------------------------------------------------
  CALCULATE PRESSURE FLOW FRICTION LOSSES(LACFCD):
  PIPE FLOW =       65.00 CFS     PIPE DIAMETER =  36.00 INCHES
  PIPE LENGTH =       100.00 FEET      MANNINGS N =  .01300
  SF=(Q/K)**2 = ((      65.00)/(    666.983))**2 =  .0094972
  HF=L*SF = (     100.00)*(  .0094972) =        .950
  NODE      4.00 : HGL= <  110.317>;EGL= <  111.630>;FLOWLINE= <   98.000>
```

```
==========================================================================
  PRESSURE FLOW PROCESS FROM NODE     4.00 TO NODE     4.00 IS CODE =  5
  UPSTREAM NODE      4.00      ELEVATION =    98.00
--------------------------------------------------------------------------
  CALCULATE PRESSURE FLOW JUNCTION LOSSES:
   NO. DISCHARGE  DIAMETER     AREA     VELOCITY    DELTA       HV
    1      25.0     24.00      3.142     7.958       .000       .983
    2      65.0     36.00      7.069     9.196        --       1.313
    3      10.0     18.00      1.767     5.659     45.000        -
    4      30.0     24.00      3.142     9.549     45.000        -
    5        .0===Q5 EQUALS BASIN INPUT===

  LACFCD AND OCEMA PRESSURE FLOW JUNCTION FORMULAE USED:
  DY=(Q2*V2-Q1*V1*COS(DELTA1)-Q3*V3*COS(DELTA3)-
      Q4*V4*COS(DELTA4))/((A1+A2)*16.1)

  UPSTREAM MANNINGS N =  .01300
  DOWNSTREAM MANNINGS N =  .01300
  UPSTREAM FRICTION SLOPE =  .01221
  DOWNSTREAM FRICTION SLOPE =  .00950
  AVERAGED FRICTION SLOPE IN JUNCTION ASSUMED AS  .01085
  JUNCTION LENGTH(FEET) =   5.00      FRICTION LOSS =    .054
  ENTRANCE LOSSES =    .000
  JUNCTION LOSSES = DY+HV1-HV2+(FRICTION LOSS)+(ENTRANCE LOSSES)
  JUNCTION LOSSES =    .950+  .983- 1.313+(  .054)+(  .000) =   .675
  NODE      4.00 : HGL= <  111.321>;EGL= <  112.304>;FLOWLINE= <   98.000>


==========================================================================
  PRESSURE FLOW PROCESS FROM NODE     4.00 TO NODE     3.00 IS CODE =  1
  UPSTREAM NODE      3.00      ELEVATION =    98.50
--------------------------------------------------------------------------
  CALCULATE PRESSURE FLOW FRICTION LOSSES(LACFCD):
  PIPE FLOW =       25.00 CFS      PIPE DIAMETER =  24.00 INCHES
  PIPE LENGTH =        60.00 FEET      MANNINGS N =  .01300
  SF=(Q/K)**2 = ((      25.00)/(    226.224))**2 =  .0122125
  HF=L*SF = (       60.00)*(  .0122125) =       .733
  NODE      3.00 : HGL= <  112.054>;EGL= <  113.037>;FLOWLINE= <   98.500>


==========================================================================
  PRESSURE FLOW PROCESS FROM NODE     3.00 TO NODE     3.00 IS CODE =  4
  UPSTREAM NODE      3.00      ELEVATION =    98.50
--------------------------------------------------------------------------
  CALCULATE PRESSURE FLOW SUDDEN PIPE ENLARGEMENT LOSSES(LACRD):
  PIPE FLOW =       25.00 CFS
  DOWNSTREAM PIPE DIAMETER =  24.00 INCHES
  UPSTREAM PIPE DIAMETER =  18.00 INCHES
  UPSTREAM PRESSURE FLOW VELOCITY =  14.15 FEET/SEC.
  DOWNSTREAM PRESSURE FLOW VELOCITY =   7.96 FEET/SEC.
  SUDDEN PIPE-FLOW ENLARGEMENT LOSSES =
  ((V1-V2)**2)/64.4 = (( 14.147-  7.958)**2/64.4 =    .595
  NODE      3.00 : HGL= <  110.524>;EGL= <  113.632>;FLOWLINE= <   98.500>


==========================================================================
  PRESSURE FLOW PROCESS FROM NODE     3.00 TO NODE     2.00 IS CODE =  1
  UPSTREAM NODE      2.00      ELEVATION =    99.00
--------------------------------------------------------------------------
  CALCULATE PRESSURE FLOW FRICTION LOSSES(LACFCD):
  PIPE FLOW =       25.00 CFS      PIPE DIAMETER =  18.00 INCHES
  PIPE LENGTH =        40.00 FEET      MANNINGS N =  .01300
  SF=(Q/K)**2 = ((      25.00)/(    105.043))**2 =  .0566427
  HF=L*SF = (       40.00)*(  .0566427) =      2.266
  NODE      2.00 : HGL= <  112.790>;EGL= <  115.898>;FLOWLINE= <   99.000>
```

```
==============================================================================
   PRESSURE FLOW PROCESS FROM NODE     2.00 TO NODE     2.00 IS CODE =  6
   UPSTREAM NODE     2.00     ELEVATION =    99.00
------------------------------------------------------------------------------
   CALCULATE PRESSURE FLOW ANGLE-POINT LOSSES(LACRD):
   PIPE FLOW =       25.00 CFS     PIPE DIAMETER =  18.00 INCHES
   PIPE ANGLE POINT DELTA =  45.00 DEGREES
   PRESSURE FLOW ANGLE-POINT COEFFICIENT KA =  .2340
   PRESSURE FLOW VELOCITY =  14.15 FEET/SEC.
   VELOCITY HEAD =  3.108
   HAPT=KA*(VELOCITY HEAD) = ( .2340)*( 3.108) =   .727
   NODE     2.00 : HGL= <  113.517>;EGL= <  116.625>;FLOWLINE= <    99.000>

==============================================================================
   PRESSURE FLOW PROCESS FROM NODE     2.00 TO NODE     1.50 IS CODE =  1
   UPSTREAM NODE     1.50     ELEVATION =    99.50
------------------------------------------------------------------------------
   CALCULATE PRESSURE FLOW FRICTION LOSSES(LACFCD):
   PIPE FLOW =       25.00 CFS     PIPE DIAMETER =  18.00 INCHES
   PIPE LENGTH =       60.00 FEET     MANNINGS N =  .01300
   SF=(Q/K)**2 = ((      25.00)/(   105.043))**2 =  .0566427
   HF=L*SF = (      60.00)*(  .0566427) =     3.399
   NODE     1.50 : HGL= <  116.916>;EGL= <  120.023>;FLOWLINE= <    99.500>

==============================================================================
   PRESSURE FLOW PROCESS FROM NODE     1.50 TO NODE     1.50 IS CODE =  9
   UPSTREAM NODE     1.50     ELEVATION =    99.50
------------------------------------------------------------------------------
   CALCULATE PRESSURE FLOW TRANSITION LOSSES(LACFCD):
   PRESSURE PIPE FLOW(CFS) =         25.00
   UPSTREAM DIAMETER(INCH) =  15.00     DOWNSTREAM DIAMETER(INCH) =  18.00
   TRANSITION STRUCTURE LENGTH(FEET) =    5.00
   UPSTREAM MANNINGS N =  .0130     DOWNSTREAM MANNINGS N =  .0130
   TOTAL-ANGLE-OF-TRANSITION(DEGREES) =  1.430
   UPSTREAM VELOCITY(FPS) =  20.372   UPSTREAM VELOCITY HEAD =  6.444
   DOWNSTREAM VELOCITY(FPS) =  14.147   DOWNSTREAM VELOCITY HEAD =  3.108
   HT = .2*(V1*V1-V2*V2)/64.4
   TRANSITION LOSS COEFFICIENT =  .2000
   TRANSITION LOSS =   .667
   UPSTREAM FRICTION SLOPE =  .14978
   DOWNSTREAM FRICTION SLOPE =  .05664
   AVERAGE TRANSITION STRUCTURE FRICTION SLOPE ASSUMED =  .10321
   HF = L*SF =   .516
   TOTAL LOSSES = (HT)+(HF) = (  .667)+(  .516)
   NODE     1.50 : HGL= <  114.762>;EGL= <  121.207>;FLOWLINE= <    99.500>

==============================================================================
   PRESSURE FLOW PROCESS FROM NODE     1.50 TO NODE     1.00 IS CODE =  1
   UPSTREAM NODE     1.00     ELEVATION =   100.00
------------------------------------------------------------------------------
   CALCULATE PRESSURE FLOW FRICTION LOSSES(LACFCD):
   PIPE FLOW =       25.00 CFS     PIPE DIAMETER =  15.00 INCHES
   PIPE LENGTH =       40.00 FEET     MANNINGS N =  .01300
   SF=(Q/K)**2 = ((      25.00)/(    64.598))**2 =  .1497765
   HF=L*SF = (      40.00)*(  .1497755) =     5.991
   NODE     1.00 : HGL= <  120.753>;EGL= <  127.198>;FLOWLINE= <   100.000>

==============================================================================
   PRESSURE FLOW PROCESS FROM NODE     1.00 TO NODE     1.00 IS CODE =  8
   UPSTREAM NODE     1.00     ELEVATION =   100.00
------------------------------------------------------------------------------
```

```
   CALCULATE PRESSURE FLOW CATCH BASIN ENTRANCE LOSSES(LACFCD):
   PIPE FLOW(CFS) =       25.00      PIPE DIAMETER(INCH) =  15.00
   PRESSURE FLOW VELOCITY HEAD =  6.444
   CATCH BASIN ENERGY LOSS = .2*(VELOCITY HEAD) = .2*( 6.444) = 1.289
   NODE     1.00 : HGL= <  128.486>;EGL= <  128.486>;FLOWLINE= <  100.000>
================================================================================
   END OF PRESSURE FLOW HYDRAULICS PIPE SYSTEM
```

CHAPTER 3.

HYDRAULIC ELEMENTS I

Introduction

In this chapter, several complete computer programs will be presented which provide the computer capability of solving for the various channel flow dimensions associated with the normal depth. Because it is preferable to link together all of the programs presented in this chapter, PROGRAM 11 provides a MAIN MENU title page wherefrom the user may select the desired computational routine. For normal depth flow in prismatic channels, PROGRAM 12 solves for any one of the variables: channel base, channel side slope or z-factor, flow depth, or flow rate given the remaining variables. Also included in PROGRAM 12 is the analysis of critical depth flow variables. Pipeflow normal and critical depth flow analysis is provided in PROGRAM 13. PROGRAM 14 provides for the normal depth flow analysis of a symmetrical roadway section. The engineer enters values for the street halfwidth (the street section is assumed to be symmetrical about the roadway centerline), and dimensions of the gutter including curb depth, gutter lip and gutter hike. Given the specified flow rate, the streetflow analysis considers both symmetrical flow and split flow effects using Manning's equation with a friction factor set at n = 0.015.

A powerful analysis routine for the study of open channel pipeflow effects at a junction point is provided by PROGRAM 15. Given an upstream and downstream pipe section, one or two lateral drains into the junction, the drop in elevation through the junction, flow rates and friction factors for each pipeline, and assumed hydraulic controls, PROGRAM 15 analyzes the junction hydraulics for the determination of the true hydraulic control and then computes the appropriate upstream and downstream pipeline flow depths. The method of analysis used is based on the Thompson equation (see Chapter 2) which relates the change in HGL to the pressure plus momentum variation through the junction with minor losses ignored. The program uses an iterative procedure to balance the pressure plus momentum relationships.

PROGRAM 16 provides the computer capability for the gradually varied flow analysis of trapezoidal channels. Similarly, PROGRAM 17 provides for the gradually varied flow analysis of pipelines. Consequently, the complete water surface profile for pipeline storm drains can be studied by using PROGRAM numbers 15 and 17, with the engineer determining the location and length of hydraulic jumps where appropriate.

PROGRAM 11. Hydraulic Elements Main Menu

The necessary software and data entry sequence to formulate a package program is presented. The utility of this program is that PROGRAM 12 through PROGRAM 17 are combined into a single program system.

PROGRAM 12. Channel Hydraulic Elements

Normal depth flow in regular open channels can be analyzed by use of Manning's equation

$$Q = 1.486 \times A \times R^{.667} \times S_f^{.5}/n \qquad (3.1)$$

where

Q = flowrate, in cfs
A = area of flow, in square feet
R = flow area divided by the wetted perimeter
S_f = friction slope
n = friction factor

An unknown value for one of the variables used in Eq. (3.1) can be estimated by use of PROGRAM 12. The computer code is arranged for estimating the selected unknown variable by means of an iteration procedure until the unknown value is within a programmed tolerance of satisfying Manning's equation. Also included in the program is the computation of critical depth hydraulic information corresponding to the normal depth flow computed in Eq. (3.1).

PROGRAM 13. Pipeflow Hydraulic Elements

Normal depth and critical depth estimation for pipeflow is provided by PROGRAM 13. Also included is the estimation of pipe capacity flows, friction slopes, and the computation of other hydraulic factors such as specific energy, Froude number, pressure plus momentum, and flow velocity. The normal depth estimates correspond to the smaller value when flow depths exceed 0.82 × pipe diameter.

PROGRAM 14. Streetflow Hydraulic Elements

The analysis of streetflow is a commonly occurring problem in civil engineering design. In this program, the engineer can analyze the special case of capacity flow on one side of the street section with the remaining flows occurring on the other side of the street. The streetflow hydraulics are analyzed by means of Manning's equation for normal depth flow. The street section is subdivided into two subreaches where one channel spans from the street centerline to the edge of the street gutter, and the second channel spans from the edge of the gutter to the inside edge of the street curb. All flows outside of the street curb are assumed to be in a ponded condition with a negligible contribution to the total flow capacity of the system, (see Fig. 3.1).

The street section is assumed to be symmetrical about the street centerline (consequently, data entry corresponds to both sides of the street section). The engineer enters the street curb height, gutter width, gutter hike (rise in elevation along the gutter width), gutter lip (difference in elevation between roadway section and top of gutter), street half width, and roadway crossfall (slope of the pavement to the gutter). Given a flowrate, PROGRAM 14 computes the normal depth flow hydraulics for either the case of symmetrical streetflow or split flow.

PROGRAM 15. Pipeflow Junction Analysis

A difficult problem which occurs in pipeflow hydraulics is the analysis of a junction where open channel flow effects need to be considered. PROGRAM 15 provides the computer capability to determine the hydraulic control at the junction and the corresponding flow depths at the downstream and upstream points of the junction.

The analysis of the junction hydraulics is by means of the Thompson equation (Chapter 2) which relates pipeflow pressure plus momentum to the change in the hydraulic grade line through the junction. In the case of open channel flow, however, the hydraulic parameters of flow depth, flow area, velocity, pressure plus momentum, and friction slope are all computational variables. In order to achieve a pressure plus momentum balance, an iterative procedure is used which sequentially halves the differences in flow depth until the Thompson equation is satisfied.

An examination of the computer code reveals that several tests are performed in order to determine the true hydraulic control. For example, the pressure plus momentum function is used to determine whether a hydraulic jump occurs at or near the junction, or whether sufficient momentum exists in the junction inflows to wash the hydraulic jump downstream. Additionally, it is noted that minor losses are ignored in the balance of pressure plus momentum. However, the computer code can be modified to include the various losses required by local agencies for inclusion in the junction analysis.

The engineer enters data values of pipesizes, friction factors, pipe slopes, drop in elevation through the junction structure, angle of approach with respect to the downstream mainline pipe, flowrates, and estimates of the downstream and upstream flow depths, (see example). The program considers the entered flowdepths in the computation of the hydraulic control. Should computed flowdepths exceed 0.82 × pipe diameter, then the

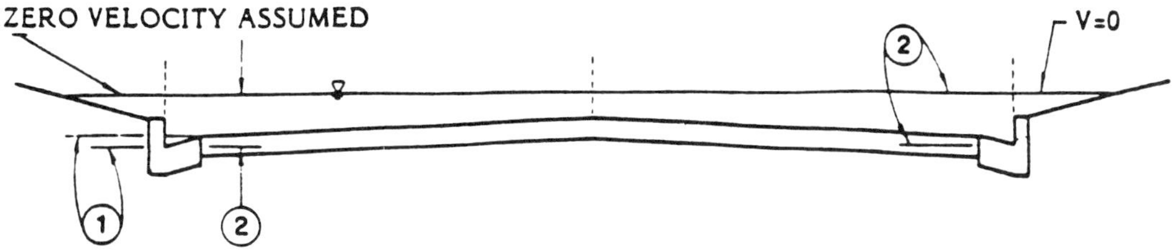

$$Q_j = \left(\frac{1.486}{.015}\right) A_j (A_j/WP_j)^{2/3} S^{1/2}$$

Where

A_1	=	GUTTER (LIP+HIKE+LIP) /2
WP_1	=	2(LIP) + HIKE + GUTTER
A_2	=	GUTTER (Y-HIKE/2)+1/2(Y-HIKE-LIP)^2/XFALL
WP_2	=	Y^1+GUTTER+LIP+(Y-HIKE-LIP)/XFALL

Where

Y^1	=	Y, for Y less than CURB, CURB, otherwise

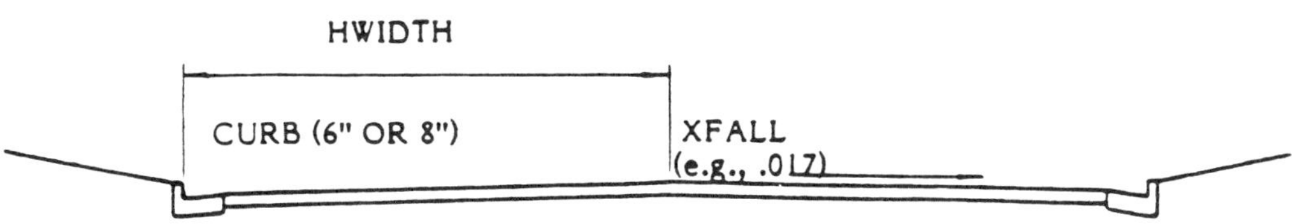

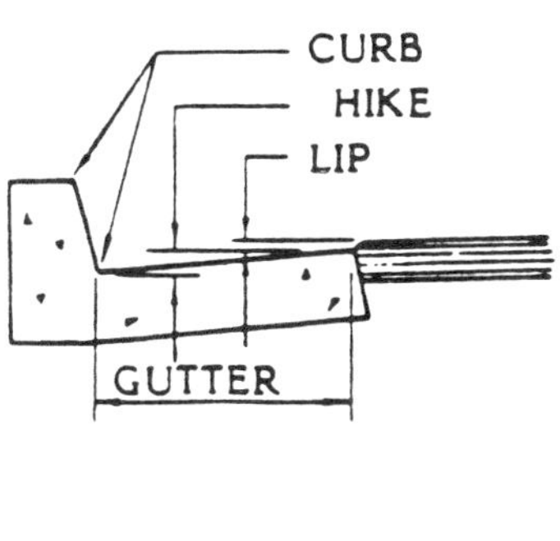

Fig. 3.1. Street flow model flow approximations

program terminates and notifies the engineer to consider a pressure flow analysis (see PROGRAM 6). Downstream pressure flow conditions may be considered in the program by entering the difference between the downstream pipe flowline and the assumed HGL as the downstream flowdepth. This type of analysis may be important in cases where sufficient inflow pressure plus momentum may force a hydraulic jump downstream of the junction structure.

PROGRAM 16. Gradually Varied Flow in Open Channels

The determination of backwater curves, drawdown curves, and other gradually varied flow profiles is an important analysis problem in open channel flow hydraulics. PROGRAM 16 provides a computer capability for the estimation of such water surface profiles for the case of rectangular, trapezoidal, and V-shaped channels. The program determines the normal and critical depth, the hydraulic control, the governing water surface profile classification, and the water surface profile. Flow depth, velocity, specific energy, and pressure plus momentum are included in the computation results.

PROGRAM 17. Gradually Varied Flow in Pipelines

Similar to PROGRAM 16, this program provides the computer capability for the analysis of gradually varied flow in pipelines. Included in the program results is the determination of normal and critical depth, the hydraulic control, water surface profile, specific energy, pressure plus momentum, and flow depths.

It is noted that a combination of PROGRAM 15 and 17 provides a complete computer capability for the analysis of open channel flow hydraulics in pipelines. However, the determination of the location and length of hydraulic jumps is not included in the programming. Rather, this type of information is currently indeterminate and is left to the engineer for special consideration on a case by case basis. A common approach is to assume the jump to occur as a shock whereby the conjugate depths are matched at a single point, with the length of the jump being assumed as zero. This type of solution may be unacceptable in cases where a pipe lateral enters the main channel immediately upstream of such an assumed hydraulic jump shock, and the hydraulic control for the pipeline is assumed to be the lower conjugate depth. Because of the similarity between PROGRAM 17 and PROGRAM 16, the data entry sequences are combined in the provided screen pages.

PROGRAM 11: DATA ENTRY

```
                    ---M A I N   M E N U---
HYDRAULIC PROCESSES:

  NON-PRESSURE FLOW ANALYSIS OF:
    1= Rectangular channel
    2= Trapezoidal channel
    3= (V) -Shaped channel
    4= Pipe
    5= Symmetrical street
    6= Pipe-flow junction

  GRADUALLY VARIED FLOW IN:
    7= Rectangular channel
    8= Trapezoidal channel
    9= (V) -Shaped channel
   10= Pipe

Select hydraulic process desired.................... ===>    "KTYPE"



--------------------------------------------------------------------------------
    TYPE: EXIT to leave program ; TOP  to go to top of page
```

PROGRAM 11

```
C  -----------------------------------------------------------------------
         PROGRAM HELEI
C  -----------------------------------------------------------------------
C
C  EXECUTIVE DRIVER FOR HELE I BATCH SYSTEM
C
         COMMON/NUT/NUT
C
C
         NUT=6
C
C
C  SELECT HYDRAULIC PROCESS
C
C  PROCESSES
100      READ FREE(5)KTYPE
         IF(KTYPE.EQ.99)GO TO 1000
C
C  TRANSLATE TO ORGINAL KTYPE IN VERSION 1.0
         IF(KTYPE.LT.7)GO TO 105
         KTYPE=KTYPE-2
         GO TO 110
105      IF(KTYPE.NE.5)GO TO 106
         KTYPE=9
         GO TO 110
106      IF(KTYPE.NE.6)GO TO 110
         KTYPE=10
110      CONTINUE
C
         GO TO(300,300,300,310,320,320,320,320,340,351),KTYPE
C
300      CALL TRAP(KTYPE)
         GO TO 100
310      CALL PIPE
         GO TO 100
320      CALL GVF(KTYPE)
         GO TO 100
340      CALL STREET
         GO TO 100
351      CALL OPCHAN
C
1000     CONTINUE
C
         END
```

PROGRAM 12: DATA ENTRY

```
---DATA ENTRY FOR RECTANGULAR CHANNEL HYDRAULICS---PAGE 1


NORMAL-DEPTH FLOW UNKNOWN VARIABLE OPTIONS:
   1= Depth(FEET)
   2= Flow(CFS)
   3= Channel slope(FEET/FEET)
   4= Basewidth(FEET)

Select desired UNKNOWN variable option number....... ===>    "KVAR"


------------------------------------------------------------------------------
TYPE: EXIT to leave program ; TOP  to go to top of page
      MAIN to go to main menu
```

```
---DATA ENTRY FOR RECTANGULAR CHANNEL HYDRAULICS---PAGE 2

Enter channel normal-depth(FEET).................... ===>    "DN"
:ALLOWABLE VALUES ARE [.01    ] TO [200   ]



Enter channel base(FEET)............................ ===>    "B"
:ALLOWABLE VALUES ARE [.1     ] TO [1000  ]

Enter channel slope(FEET/FEET)...................... ===>    "S"
:ALLOWABLE VALUES ARE [.00001] TO [.25   ]

Enter channel flow(CFS)............................. ===>    "Q"
:ALLOWABLE VALUES ARE [0      ] TO [1000000   ]


------------------------------------------------------------------------------
TYPE: EXIT to leave program ; TOP  to go to top of page
                            ; BACK to go back one page
```

```
---DATA ENTRY FOR TRAPEZOIDAL CHANNEL HYDRAULICS---PAGE 1



NORMAL-DEPTH FLOW UNKNOWN VARIABLE OPTIONS:
   1= Depth(FEET)
   2= Flow(CFS)
   3= Channel slope(FEET/FEET)
   4= Basewidth(FEET)
   5= Channel Z (QUOTIENT OF HORIZONTAL/VERTICAL)

Select desired UNKNOWN variable option number....... ===>    "KVAR"



--------------------------------------------------------------------------------
TYPE: EXIT to leave program ; TOP  to go to top of page
      MAIN to go to main menu
```

```
---DATA ENTRY FOR TRAPEZOIDAL CHANNEL HYDRAULICS---PAGE 2

Enter channel normal-depth(FEET).................... ===>    "DN"
:ALLOWABLE VALUES ARE [.01    ] TO [200   ]

Enter channel Z value............................... ===>    "Z"
(NOTE: Z = QUOTIENT OF (HORIZONTAL)/(VERTICAL))
:ALLOWABLE VALUES ARE [0] TO [100   ]

Enter channel base(FEET)............................ ===>    "B"
:ALLOWABLE VALUES ARE [.1     ] TO [1000  ]

Enter channel slope(FEET/FEET)...................... ===>    "S"
:ALLOWABLE VALUES ARE [.00001] TO [.25   ]

Enter channel flow(CFS)............................. ===>    "Q"
:ALLOWABLE VALUES ARE [0      ] TO [1000000   ]



--------------------------------------------------------------------------------
TYPE: EXIT to leave program ; TOP  to go to top of page
                            ; BACK to go back one page
```

```
---DATA ENTRY FOR (V) -SHAPED CHANNEL HYDRAULICS---PAGE 1



NORMAL-DEPTH FLOW UNKNOWN VARIABLE OPTIONS:
   1= Depth(FEET)
   2= Flow(CFS)
   3= Channel slope(FEET/FEET)

   5= Channel Z (QUOTIENT OF HORIZONTAL/VERTICAL)

Select desired UNKNOWN variable option number....... ===>    "KVAR"







--------------------------------------------------------------------------------
    TYPE: EXIT to leave program ; TOP  to go to top of page
          MAIN to go to main menu
```

```
---DATA ENTRY FOR (V) -SHAPED CHANNEL HYDRAULICS---PAGE 2

Enter channel normal-depth(FEET).................... ===>    "DN"
ALLOWABLE VALUES ARE [.01    ] TO [200   ]

Enter channel Z value............................... ===>    "Z"
(NOTE: Z = QUOTIENT OF (HORIZONTAL)/(VERTICAL))
:ALLOWABLE VALUES ARE [0] TO [100   ]




Enter channel slope(FEET/FEET)...................... ===>    "S"
:ALLOWABLE VALUES ARE [.00001] TO [.25   ]

Enter channel flow(CFS)............................. ===>    "Q"
:ALLOWABLE VALUES ARE [0     ] TO [1000000   ]




--------------------------------------------------------------------------------
    TYPE: EXIT to leave program ; TOP  to go to top of page
                                ; BACK to go back one page
```

```
---DATA ENTRY FOR RECT/TRAP/V-SHAPED CHANNEL HYDRAULICS---PAGE 3

Enter channel friction factor(Mannings)............. ===>    "RN"
:ALLOWABLE VALUES ARE [.008  ] TO [.9999 ]
NOTE: SOME SUGGESTED VALUES ARE AS FOLLOWS:

     VALUE    CHANNEL TYPE

      .014    RECTANGULAR
      .015    TRAPEZOIDAL
      .017    ASPHALT CONCRETE
      .03     ENGINEERING EARTH, SIZE  DETERMINATION
      .02     ENGINEERING EARTH, SCOUR DETERMINATION
      .035    ROCK SLOPE PROTECTION LEVEE RIPRAP

--------------------------------------------------------------------------------
TYPE: EXIT to leave program ; TOP  to go to top of page
                            ; BACK to go back one page
```

PROGRAM 12

```
C  ---------------------------------------------------------------------
         SUBROUTINE TRAP(KTYPE)
C  ---------------------------------------------------------------------
C
C  KTYPE=1,2,3 = RECTANGULAR,TRAPEZOIDAL,V-CHANNEL
C
C  SUBROUTINE ANALYSIS UNIFORM FLOW IN A TRAPEZOIDAL CHANNEL
C
         COMMON /NUT/NUT
C
C
         YCRIT(YY)=1.-Q*Q*(B+2.*Z*YY)/(G*XAREA(YY)**3.)
         XAREA(YY)=YY*(B+Z*YY)
         FPM(YY)=GAM*(YY*YY*(B*.5+Z*YY/3.)+Q*Q/(G*XAREA(YY)))
         FN(QQ,BB,DD,ZZ,SS)=1.-QQ*RN*(BB+2.*DD*SQRT(ZZ*ZZ+1.))**.66667/
         (1.486*((BB+ZZ*DD)*DD)**1.6667*SQRT(SS))
C
         NT=NUT
C
C
C INITIALIZE REQUIRED INPUT VARIABLES
         Z=0.
         DN=0.
         B=0.
         S=0.
         Q=0.
C
C
C..READ DATA INPUT
         READ FREE(5)KVAR,DN,Z,B,S,Q,RN
C
C
         DMAX=2000.
         G=32.2
         GAM=62.4
C-------CHANNEL COMPUTATION ALGORITHMS
         WRITE(NT,187)
         WRITE(NT,185)
185      FORMAT(1X,'>>>>CHANNEL INPUT INFORMATION<<<<')
         WRITE(NT,190)
190      FORMAT(4('-------------------'))
         IF(KVAR.NE.1)WRITE(NT,191)DN
         IF(KVAR.NE.5)WRITE(NT,195)Z
         IF(KVAR.NE.4)WRITE(NT,194)B
         IF(KVAR.NE.3)WRITE(NT,193)S
         IF(KVAR.NE.2)WRITE(NT,192)Q
         WRITE(NT,196)RN
191      FORMAT(5X,'NORMAL DEPTH(FEET) = ',F6.2)
192      FORMAT(5X,'UNIFORM FLOW(CFS) = ',F11.2)
193      FORMAT(5X,'CONSTANT CHANNEL SLOPE(FEET/FEET) = ',F8.6)
194      FORMAT(5X,'BASEWIDTH(FEET) = ',F8.2)
195      FORMAT(5X,'CHANNEL Z(HORIZONTAL/VERTICAL) = ',F7.2)
196      FORMAT(5X,'MANNINGS FRICTION FACTOR = ',F6.4)
         IF(KVAR.NE.2.AND.KVAR.NE.3)GOTO 197
C
C        FIND Q OR S
C
         AREA=XAREA(DN)
         WP=B+DN*2.*SQRT(1+Z*Z)
```

```
      RH=AREA/WP
      XK=1.486*AREA*RH**.66667/RN
      IF(KVAR.EQ.2)Q=XK*SQRT(S)
      IF(XK.LE..001)WRITE(NT,2001)
2001  FORMAT(5X,'==>CHANNEL CONVEYANCE FACTOR IS LESS THAN [.001]')
      IF(XK.LE..001)GOTO 1000
      IF(KVAR.EQ.3)S=Q/XK*Q/XK
      GOTO 450
197   IF(Q.GT.0.)GOTO 200
      WRITE(NT,198)
198   FORMAT(/,3X,'ZERO FLOW SPECIFIED IN CHANNEL.',/)
      WRITE(NT,180)
      GOTO 1000
200   CONTINUE
C--------NONZERO FLOW CALCULATIONS
C
C
C     ARBITRARY CONSTANTS
C
C
      IF(B.LE.0.)B=.0001
      YMAX=DMAX
      YMIN=0.
      BMAX=1000.
      BMIN=.001
      ZMAX=100.
      ZMIN=0.
      IF(KVAR.EQ.1)GOTO 2004
      YMAX=DN
      YMIN=DN
2004  IF(KVAR.EQ.4)GOTO 2002
      BMAX=B
      BMIN=B
2002  IF(KVAR.EQ.5)GOTO 2003
      ZMAX=Z
      ZMIN=Z
C
C     CHECK FOR FUNCTIONAL DEFINITIONS
C
C
2003  IF(KVAR.EQ.4)GOTO 2020
      IF(KVAR.EQ.5)GOTO 2030
2010  F=FN(Q,B,DMAX,Z,S)
      IF(F.GE.0.)GOTO 210
2013  WRITE(NT,180)
      IF(F.LT.0.)WRITE(NT,202)
202   FORMAT(3X,'FLOW EXCEEDS CHANNEL CAPACITY.',/)
      WRITE(NT,180)
      GOTO 1000
2020  F=FN(Q,BMIN,DN,Z,S)
      IF(F.LT.0.)GOTO 2025
      WRITE(NT,180)
      WRITE(NT,2021)
2021  FORMAT(5X,'NORMAL DEPTH FLOW WITH BASEWIDTH AT [0] EXCEEDS'
     C ' DESIRED CHANNEL FLOW.')
      WRITE(NT,180)
      GOTO 1000
2025  F=FN(Q,BMAX,DN,Z,S)
      IF(F.GE.0.)GOTO 210
      WRITE(NT,180)
```

```
      WRITE(NT,2026)
2026  FORMAT(5X,'FLOW EXCEEDS CHANNEL CAPACITY WITH BASE(FEET) = [1000]')
      WRITE(NT,180)
      GOTO 1000
2030  F=FN(Q,B,DN,ZMIN,S)
      IF(F.LT.0.)GOTO 2035
      WRITE(NT,180)
      WRITE(NT,2031)
2031  FORMAT(5X,'NORMAL FLOW WITH Z = [0] EXCEEDS DESIRED CHANNEL FLOW.')
      WRITE(NT,180)
      GOTO 1000
2035  F=FN(Q,B,DN,ZMAX,S)
      IF(F.GE.0.)GOTO 210
      WRITE(NT,180)
      WRITE(NT,2036)
2036  FORMAT(5X,'FLOW EXCEEDS CHANNEL CAPACITY WITH Z = [100]')
      WRITE(NT,180)
      GOTO 1000
210   DO 440 I=1,20
      DN=.5*(YMIN+YMAX)
      Z=.5*(ZMIN+ZMAX)
      B=.5*(BMIN+BMAX)
      F=FN(Q,B,DN,Z,S)
      IF(F)420,450,430
420   YMIN=DN
      BMIN=B
      ZMIN=Z
      GOTO 440
430   YMAX=DN
      BMAX=B
      ZMAX=Z
440   CONTINUE
450   CONTINUE
      TW=B+2.*Z*DN
      AREA=.5*(B+TW)*DN
      V=Q/AREA
      DH=AREA/TW
      FR=V/SQRT(32.2*DH)
      FPMDN=FPM(DN)
C
C
      WRITE(NT,180)
      WRITE(NT,490)
490   FORMAT(5X,'NORMAL-DEPTH FLOW INFORMATION:')
      WRITE(NT,190)
      IF(KVAR.EQ.1)WRITE(NT,495)DN
      IF(KVAR.EQ.5)WRITE(NT,497)Z
      IF(KVAR.EQ.4)WRITE(NT,496)B
      IF(KVAR.EQ.3)WRITE(NT,499)S
      IF(KVAR.EQ.2)WRITE(NT,498)Q
495   FORMAT(5X,'>>>>> NORMAL DEPTH(FEET) = ',F6.2)
496   FORMAT(5X,'>>>>> BASEWIDTH(FEET) = ',F7.2)
497   FORMAT(5X,'>>>>> CHANNEL Z-FACTOR = ',F6.2)
498   FORMAT(5X,'>>>>> NORMAL DEPTH FLOW(CFS) = ',F9.2)
499   FORMAT(5X,'>>>>> CHANNEL SLOPE(FEET/FEET) = ',F8.5)
      WRITE(NT,500)TW,AREA,DH,V,FR,FPMDN
500   FORMAT(5X,'FLOW TOP- WIDTH(FEET) = ',F12.2,/,
     C 5X,'FLOW AREA(SQUARE FEET) = ',F14.2,/,
     C 5X,'HYDRAULIC DEPTH(FEET) = ',F8.2,/,
     C 5X,'FLOW AVERAGE VELOCITY(FEET/SEC.) = ',F8.2,/,
```

```
     C  5X,'UNIFORM FROUDE NUMBER = ',F8.3,/,
     C  5X,'PRESSURE + MOMENTUM(POUNDS) = ',F15.2)
        HV=V*V/64.4
        ENERGY=DN+HV
        WRITE(NT,502)HV,ENERGY
502     FORMAT(5X,'AVERAGED VELOCITY HEAD(FEET) = ',F10.3,/,
     C  5X,'SPECIFI: ENERGY(FEET) = ',F10.3)
600     CONTINUE
C-------CRITICAL DEPTH COMPUTATIONS
        UP=10000.
        DN=0.
        YC=UP/2.
        DO 620 J=1,22
        IF(YCRIT(YC))614,621,615
614     DN=YC
        GOTO 620
615     UP=YC
620     YC=(UP+DN)*.5
621     AREA=XAREA(YC)
        V=Q/AREA
        TW=B+2.*YC*Z
        DH=AREA/TW
        FPMDC=FPM(YC)
C
        WRITE(NT,180)
        WRITE(NT,630)
630     FORMAT(5X,'CRITICAL-DEPTH FLOW INFORMATION:')
        WRITE(NT,190)
        WRITE(NT,640)TW,AREA,DH,V,YC,FPMDC
640     FORMAT(5X,'CRITICAL FLOW TOP-WIDTH(FEET) = ',F12.2,/,
     C  5X,'CRITICAL FLOW AREA(SQUARE FEET) = ',F14.2,/,
     C  5X,'CRITICAL FLOW HYDRAULIC DEPTH(FEET) = ',F8.2,/,
     C  5X,'CRITICAL FLOW AVERAGE VELOCITY(FEET/SEC.) = ',F8.2,/,
     C  5X,'CRITICAL DEPTH(FEET) = ',F8.2,/,
     C  5X,'CRITICAL FLOW PRESSURE + MOMENTUM(POUNDS) = ',F15.2)
        HV=V*V/64.4
        ENERGY=HV+YC
        WRITE(NT,642)HV,ENERGY
642     FORMAT(5X,'AVERAGED CRITICAL FLOW VELOCITY HEAD(FEET) = ',F10.3,/,
     C  5X,'CRITICAL FLOW SPECIFIC ENERGY(FEET) = ',F10.3)
1000    CONTINUE
C
C
C  FORMATS
C
180     FORMAT(76('='))
187     FORMAT(76('*'))
C
        RETURN
        END
```

```
**********DESCRIPTION OF RESULTS*******************************************
*                                                                         *
*                  TRAPEZOIDAL CHANNEL FLOW INFORMATION                   *
*                                                                         *
***************************************************************************

===========================================================================
     CHANNEL INPUT INFORMATION IS AS FOLLOWS:
---------------------------------------------------------------------------
     UNIFORM FLOW(CFS) =      1000.00
     CONSTANT CHANNEL SLOPE(FEET/FEET) =  .005000
     BASEWIDTH(FEET) =      5.00
     CHANNEL Z(HORIZONTAL/VERTICAL) =     2.00
     MANNINGS FRICTION FACTOR =  .0300
===========================================================================
===========================================================================
     NORMAL-DEPTH FLOW INFORMATION:
---------------------------------------------------------------------------
     ***** NORMAL DEPTH(FEET) =   6.70
     FLOW TOP- WIDTH(FEET) =         31.80
     FLOW AREA(SQUARE FEET) =          123.30
     HYDRAULIC DEPTH(FEET) =     3.88
     FLOW AVERAGE VELOCITY(FEET/SEC.) =      8.11
     UNIFORM FROUDE NUMBER =     .726
     PRESSURE + MOMENTUM(POUNDS) =         35235.84
     AVERAGED VELOCITY HEAD(FEET) =      1.021
     SPECIFIC ENERGY(FEET) =      7.722
===========================================================================
===========================================================================
     CRITICAL-DEPTH FLOW INFORMATION:
---------------------------------------------------------------------------
     CRITICAL FLOW TOP-WIDTH(FEET) =         28.09
     CRITICAL FLOW AREA(SQUARE FEET) =           95.53
     CRITICAL FLOW HYDRAULIC DEPTH(FEET) =     3.40

     CRITICAL FLOW PRESSURE + MOMENTUM(POUNDS) =         33490.68
     AVERAGED CRITICAL FLOW VELOCITY HEAD(FEET) =      1.702
     CRITICAL FLOW SPECIFIC ENERGY(FEET) =      7.475
===========================================================================
```

PROGRAM 13: DATA ENTRY

```
---DATA ENTRY FOR PIPE FLOW HYDRAULICS---PAGE 1



NORMAL-DEPTH FLOW UNKNOWN VARIABLE OPTIONS:
   1= Depth(FEET)
   2= Flow(CFS)
   3= Pipe-Capacity(soffit flow) DIAMETER(FEET)
   4= Pipe-Capacity(soffit flow) SLOPE(FEET/FEET)

Select desired UNKNOWN variable option number....... ===>    "KVAR"

--------------------------------------------------------------------------------
TYPE: EXIT to leave program ; TOP  to go to top of page
      MAIN to go to main menu
```

```
---DATA ENTRY FOR PIPE FLOW HYDRAULICS---PAGE 2

Enter constant pipe diameter(FEET).................. ===>    "DIAM"
:ALLOWABLE VALUES ARE [.1    ] TO [20    ]

Enter pipe flowdepth(FEET).......................... ===>    "YN"
:ALLOWABLE VALUES ARE [.1    ] TO [20    ]

Enter constant channel slope(FEET/FEET)............. ===>    "SO"
:ALLOWABLE VALUES ARE [.00001] TO [.99   ]

Enter constant channel flow(CFS).................... ===>    "Q"
:ALLOWABLE VALUES ARE [.01   ] TO [1000000   ]

Enter channel friction factor(Mannings)............. ===>    "RN"
:ALLOWABLE VALUES ARE [.008  ] TO [.9999 ]
NOTE: SOME SUGGESTED VALUES ARE AS FOLLOWS:
      .013    REINFORCED CONCRETE PIPE(RCP)
      .014    CAST-IN-PLACE(CIPP)
      .024    CORRUGATED STEEL PIPE(CSP)

--------------------------------------------------------------------------------
TYPE: EXIT to leave program ; TOP  to go to top of page
                            ; BACK to go back one page
```

PROGRAM 13

```
C ------------------------------------------------------------------
        SUBROUTINE PIPE
C ------------------------------------------------------------------
C
C  ANALYSIS OF PIPE FLOW
C
        COMMON /NUT/NUT
C
        DELTA(YY)=ACOS((R-YY)/R)
        AREA(YY)=R*R*(ANG-.5*SIN(2.*ANG))
        PER(YY)=2.*R*ANG
        TW(YY)=2.*R*SIN(ANG)
        YCRIT(YY)=1.-Q*Q*TW(YY)/(G*AREA(YY)**3.)
        YNORM(YY)=1.-Q*Q*CON/(AREA(YY)**3.3333/PER(YY)**1.3333)
        DL(YY)=YCRIT(YY)/(YNORM(YY)*S0)
        FPM(YY)=GAM*(Q*Q/(G*AREA(YY))+YB*AREA(YY))
        ENERGY(YY)=YY+Q*Q/(2.*G*AREA(YY)**2.)
C
        NT=NUT
C
C INITIALIZE REQUIRED INPUT VARIABLES
        YN=0.
        DIAM=0.
        B=0.
        S0=0.
        Q=0.
C
C
C
C..READ DATA INPUT
        READ FREE(5)KVAR,DIAM,YN,S0,Q,RN
C
C
C  CONSTANTS
        GAM=62.4
        G=32.2
        CON=(RN/1.486)**2./S0
        WRITE(NT,187)
6       FORMAT(76('-'))
        WRITE(NT,200)
200     FORMAT(1X,'>>>>PIPEFLOW HYDRAULIC INPUT INFORMATION<<<<')
        WRITE(NT,6)
        IF(KVAR.EQ.4)YN=DIAM
        IF(KVAR.NE.3)WRITE(NT,202)DIAM
        IF(KVAR.NE.1.AND.KVAR.NE.3)WRITE(NT,201)YN
        IF(KVAR.NE.4)WRITE(NT,203)S0
        IF(KVAR.NE.2)WRITE(NT,204)Q
        WRITE(NT,205)RN
201     FORMAT(5X,'FLOWDEPTH(FEET) = ',F7.3)
202     FORMAT(5X,'PIPE DIAMETER(FEET) = ',F7.3)
203     FORMAT(5X,'PIPE SLOPE(FEET/FEET) = ',F7.4)
204     FORMAT(5X,'PIPEFLOW(CFS) = ',F12.2)
205     FORMAT(5X,'MANNINGS FRICTION FACTOR = ',F8.6)
        R=DIAM/2.
C
C       SOLVE FOR DIAM AND Q FOR CAPACITY PROBLEMS
C
        GOTO(511,5210,2000,2500)KVAR
C.......FIND DIAM
```

```
2000    XK=.46316/RN*SQRT(S0)
        DIAM=(Q/XK)**.375
        WRITE(NT,2001)DIAM
2001    FORMAT(5X,'>>>>>SOFFIT-FLOW PIPE DIAMETER(FEET) = ',F7.3)
        GOTO 1000
2500    XK=.463164/RN*(DIAM**2.66667)
        S0=Q/XK*Q/XK
        WRITE(NT,2501)S0
2501    FORMAT(5X,'>>>>>SOFFIT-FLOW PIPE SLOPE(FEET/FEET) = ',F7.4)
        GOTO 1000
511     DMAX=.98*DIAM
512     UP=DIAM
        DN=0.
        YC=DMAX/2.
        DO 520 I=1,22
        ANG=DELTA(YC)
        IF(YCRIT(YC))514,521,515
514     DN=YC
        GOTO 520
515     UP=YC
520     YC=(UP+DN)*.5
521     CONTINUE
        AREAC=AREA(YC)
        ENGC=ENERGY(YC)
        TWC=TW(YC)
        CALL YBAR(YC,DIAM,YB,NT)
        FPMC=FPM(YC)
        VC=Q/AREAC
        HVC=VC*VC/64.4
        DHC=AREAC/TWC
        WRITE(NT,180)
        WRITE(NT,781)
781     FORMAT(5X,'CRITICAL-DEPTH FLOW INFORMATION:')
        WRITE(NT,6)
        WRITE(NT,601)YC,AREAC,TWC,FPMC,VC,HVC,DHC,ENGC
601     FORMAT(5X,'CRITICAL DEPTH(FEET) = ',F9.2,/,
     C  5X,'CRITICAL FLOW AREA(SQUARE FEET) = ',F9.3,/,
     C  5X,'CRITICAL FLOW TOP-WIDTH(FEET) = ',F7.3,/,
     C  5X,'CRITICAL FLOW PRESSURE + MOMENTUM(POUNDS) = ',F15.2,/,
     C  5X,'CRITICAL FLOW VELOCITY(FEET/SEC.) = ',F9.3,/,
     C  5X,'CRITICAL FLOW VELOCITY HEAD(FEET) = ',F15.2,/,
     C  5X,'CRITICAL FLOW HYDRAULIC DEPTH(FEET) = ',F9.2,/,
     C  5X,'CRITICAL FLOW SPECIFIC ENERGY(FEET) = ',F15.2)
C:::::::USE LOWER LIMB FOR NORMAL DEPTH
        IF(KVAR.EQ.1)GOTO 5220
C
C       FIND Q(D)
C
5210    IF(YN.GE.DIAM)KODE=1
        DM=.99*DIAM
        IF(YN.GE.DM)YN=DM
        XK=1.486/RN*SQRT(S0)
        Q=.463164/RN*SQRT(S0)*(DIAM**2.66667)
        IF(KODE.EQ.1)GOTO 777
        ANG=DELTA(YN)
        AREAC=AREA(YN)
        PERC=PER(YN)
        RH=AREAC/PERC
        Q=XK*(RH**.666667)*AREAC
777     WRITE(NT,534)Q
```

```
534     FORMAT(5X,'>>>>> NORMAL DEPTH FLOW(CFS) = ',F9.2)
        IF(KODE.EQ.1)GOTO 1000
        GOTO 535
5220    TEST=.498*DIAM**2.6667*SO**.5/RN
        IF(Q.LT.TEST)GOTO 522
        WRITE(NT,5221)
5221    FORMAT(5X,'==>NORMAL PIPEFLOW IS PRESSURE FLOW')
        YN=DIAM*2.
        GOTO 1000
522     UP=DMAX
        DN=0.
        YN=DMAX/2.
        DO 530 I=1,22
        ANG=DELTA(YN)
        IF(YNORM(YN))523,531,524
523     DN=YN
        GOTO 530
524     UP=YN
530     YN=(UP+DN)*.5
531     CONTINUE
        DMAX=.82*DIAM
        IF(YN.LT.DIAM.AND.YN.GT.DMAX)WRITE(NT,5331)
5331    FORMAT(5X,'NOTE:GIVEN NORMAL DEPTH IS LOWER VALUE OF TWO POSSIBLE.'
     C  ,/,5X,'SUGGEST CONSIDERATION OF WAVE ACTION, UNCERTAINTY, ETC.')
        AREAC=AREA(YN)
535     ENGC=ENERGY(YN)
        TWC=TW(YN)
        CALL YBAR(YN,DIAM,YB,NT)
        FPMC=FPM(YN)
        VC=Q/AREAC
        HVC=VC*VC/64.4
        DHC=AREAC/TWC
        FR=(VC*VC/32.2/DHC)**.5
C
        WRITE(NT,180)
        WRITE(NT,782)
782     FORMAT(5X,'NORMAL-DEPTH FLOW INFORMATION:')
        WRITE(NT,6)
        WRITE(NT,611)YN,AREAC,TWC,FPMC,VC,HVC,DHC,FR,ENGC
611     FORMAT(5X,'NORMAL DEPTH(FEET) = ',F9.2,/,
     C  5X,'FLOW AREA(SQUARE FEET) = ',F9.2,/,
     C  5X,'FLOW TOP WIDTH(FEET) = ',F9.3,/,
     C  5X,'FLOW PRESSURE + MOMENTUM(POUNDS) = ',F15.2,/,
     C  5X,'FLOW VELOCITY(FEET/SEC.) = ',F15.3,/,
     C  5X,'FLOW VELOCITY HEAD(FEET) = ',F15.3,/,
     C  5X,'HYDRAULIC DEPTH(FEET) = ',F9.2,/,
     C  5X,'FROUDE NUMBER = ',F9.3,/,
     C  5X,'SPECIFIC ENERGY(FEET) = ',F15.2)
1000    CONTINUE
C
C  FORMATS
C
180     FORMAT(76('='))
187     FORMAT(76('*'))
C
        RETURN
        END
C
C
C  ------------------------------------------------------------
```

```
      SUBROUTINE YBAR(Y,DIAM,YB,NUT)
C ---------------------------------------------------------------------
C
C CALCULATES CENTROID OF PIPEFLOW,YB(FEET)
C
      X=Y/DIAM
      IF(X.EQ.0.)YB=0.
      IF(X.GE.1.)YB=.5*DIAM
      IF(X.EQ.0..OR.X.GE.DIAM)GOTO 1000
      IF(X.LT..4)XK=.41666667*X
      IF(X.GE..4.AND.X.LT..55)XK=.167+(X-.4)*.45
      IF(X.GE..55.AND.X.LT..72)XK=.234+(X-.55)*.5
      IF(X.GE..72.AND.X.LT..79)XK=.320+(X-.72)*.54285
      IF(X.GE..79.AND.X.LT..88)XK=.358+(X-.79)*.5888889
      IF(X.GE..88.AND.X.LT.97)XK=.411+(X-.88)*.69667
      IF(X.GE..97)XK=.474+(X-.97)*.8666667
      YB=DIAM*XK
1000  CONTINUE
C
      RETURN
      END
```

PROGRAM 14: DATA ENTRY

```
---DATA ENTRY FOR SYMMETRICAL STREET---PAGE 1

Enter constant street grade(DECIMAL)................ ===>    "SO"
:ALLOWABLE VALUES ARE [.00001] TO [.25   ]

Enter total street flow(CFS)........................ ===>    "QQ"
:ALLOWABLE VALUES ARE [.1    ] TO [500   ]

Enter average friction factor(Manning).............. ===>    "XN"
(NOTE: RECOMMENDED VALUE IS [.015])
:ALLOWABLE VALUES ARE [.008  ] TO [.9999 ]

Enter constant symmetrical street half-width(FEET).. ===>    "HWIDTH"
:ALLOWABLE VALUES ARE [4] TO [100   ]

Enter constant symmetrical street crossfall(DECIMAL) ===>    "XFALL"
:ALLOWABLE VALUES ARE [.0001 ] TO [.5    ]

--------------------------------------------------------------------------------
TYPE: EXIT to leave program ; TOP  to go to top of page
      MAIN to go to main menu
```

```
---DATA ENTRY FOR SYMMETRICAL STREET---PAGE 2

Enter constant symmetrical street curb-height(FEET)  ===>    "CURB"
:ALLOWABLE VALUES ARE [.1    ] TO [3     ]

Enter constant symmetrical gutter-width(FEET)....... ===>    "GUTTER"
NOTE: For 6-inch curb, GUTTER-WIDTH IS USUALLY [1.5]
      For 8-inch curb, GUTTER WIDTH IS USUALLY [2.0]
:ALLOWABLE VALUES ARE [.1    ] TO [5     ]

Enter constant symmetrical gutter-lip(FEET)......... ===>    "XLIP"
NOTE: For 6-inch curb, GUTTER-LIP IS USUALLY [.03125]
      For 8-inch curb, GUTTER-LIP IS USUALLY [.03125]
:ALLOWABLE VALUES ARE [.01   ] TO [1     ]

Enter constant symmetrical gutter-hike(FEET)........ ===>    "GHIKE"
NOTE: For 6-inch curb, GUTTER-HIKE IS USUALLY [.125]
      For 8-inch curb, GUTTER-HIKE IS USUALLY [.167]
:ALLOWABLE VALUES ARE [.01   ] TO [1     ]

--------------------------------------------------------------------------------
TYPE: EXIT to leave program ; TOP  to go to top of page
                            ; BACK to go back one page
```

```
---DATA ENTRY FOR SYMMETRICAL STREET---PAGE 3




STREETFLOW OPTIONS:
    1: Runoff flows on one side of the street
    2: Runoff flows evenly on both sides of the street
Select streetflow option desired..................... ===>    "IWIDTH"




------------------------------------------------------------------------------
TYPE: EXIT to leave program ; TOP  to go to top of page
                            ; BACK to go back one page
```

PROGRAM 14

```
C ------------------------------------------------------------------
      SUBROUTINE STREET
C ------------------------------------------------------------------
C
C  ANALYSIS OF UNIFORM FLOW IN A STREET
C
      COMMON /NUT/NUT
C
      NT=NUT
C
C
C..READ DATA INPUT
      READ FREE(5)S0,QQ,XN,HWIDTH,XFALL,CURB,GUTTER,XLIP,
      GHIKE,IWIDTH
C
C
      WRITE(NT,187)
      WRITE(NT,115)
115   FORMAT(1X,'>>>>STREETFLOW MODEL INPUT INFORMATION<<<<')
      WRITE(NT,6)
      WRITE(NT,116)S0,QQ,XN,HWIDTH,XFALL
116   FORMAT(5X,'CONSTANT STREET GRADE(FEET/FEET) = ',F8.6,/,
    C 5X,'CONSTANT STREET FLOW(CFS) = ',F7.2,/,
    C 5X,'AVERAGE STREETFLOW FRICTION FACTOR(MANNING) = ',F8.6,/,
    C 5X,'CONSTANT SYMMETRICAL STREET HALF-WIDTH(FEET) = ',F7.2,/,
    C 5X,'CONSTANT SYMMETRICAL STREET CROSSFALL(DECIMAL) = ',F8.6)
      WRITE(NT,117)CURB,GUTTER,XLIP,GHIKE
117   FORMAT(5X,'CONSTANT SYMMETRICAL CURB HEIGTH(FEET) = ',F6.2,/,
    C 5X,'CONSTANT SYMMETRICAL GUTTER-WIDTH(FEET) = ',F6.2,/,
    C 5X,'CONSTANT SYMMETRICAL GUTTER-LIP(FEET) = ',F8.5,/,
    C 5X,'CONSTANT SYMMETRICAL GUTTER-HIKE(FEET) = ',F8.5)
      IF(IWIDTH.EQ.1)WRITE(NT,118)
      IF(IWIDTH.EQ.2)WRITE(NT,119)
118   FORMAT(5X,'FLOW ASSUMED TO FILL STREET ON ONE SIDE, AND THEN ',
      'SPLITS')
119   FORMAT(5X,'FLOW ASSUMED TO FILL STREET EVENLY ON BOTH SIDES')
C
      ERROR=0.
      FACTOR=SQRT(1.+XFALL*XFALL)
      ISPLIT=0
      A1=GUTTER*(XLIP+GHIKE+XLIP)/2.
      WP1=XLIP+GHIKE+SQRT(GUTTER*GUTTER+GHIKE*GHIKE)+XLIP
      HR=A1/WP1
      XK1=1.486/XN*A1*HR**.6667
      CROWN=(HWIDTH-GUTTER)*XFALL+XLIP+GHIKE
      A2=GUTTER*(CROWN-GHIKE/2.)+(HWIDTH-GUTTER)*.5*(CROWN-GHIKE-XLIP)
      WP2=WP1+CROWN-XLIP-GUTTER+(HWIDTH-GUTTER)*FACTOR
      HR=A2/WP2
      XK2=1.486/XN*A2*HR**.6667
      Q=QQ
      IF(IWIDTH.EQ.2)Q=QQ/2.
      TESTK=Q/S0**.5
      IF(TESTK.GT.XK1)GO TO 300
C-------GUTTER FLOW MODEL
      V=Q/A1
      FWIDTH=GUTTER
      DN=XLIP+GHIKE
      GO TO 3000
300   IF(TESTK.GT.XK2)GO TO 400
```

```
C-------FLOW IS LESS THAN CROWN
        YTEST1=0.
        YMAX=CROWN
        SET=XLIP+GHIKE
        YMIN=SET
        I=0
350     I=I+1
        YTEST=(YMAX+YMIN)*.5
        A=GUTTER*(YTEST-GHIKE/2.)+((YTEST-SET)**2.)/XFALL/2.
        WP=WP1+YTEST-SET+((YTEST-SET)/XFALL)*FACTOR
        HR=A/WP
        XK=1.486/XN*A*HR**.6667
        TEST=TESTK-XK
        IF(TEST)360,380,365
360     YMAX=YTEST
        GO TO 370
365     YMIN=YTEST
370     TEST=ABS(YTEST1-YTEST)
        IF(TEST.LT..01)GO TO 380
        IF(I.GT.50)GO TO 2900
        YTEST1=YTEST
        GO TO 350
380     V=Q/A
        DN=YTEST
        FWIDTH=GUTTER+(DN-SET)/XFALL
        IF(ISPLIT.EQ.1)GOTO 437
        GO TO 3000
400     CONTINUE
C-------FLOW EXCEEDS CROWN
        IF(IWIDTH.EQ.2)GO TO 440
        CFULLK=2.*XK2
        IF(TESTK.LT.CFULLK)GO TO 430
        Q=Q/2.
        TESTK=Q/S0**.5
        GO TO 440
430     DN=CROWN
C-------FLOW SPLITS AND IS LESS THAN FULL(CROWN) STREET
        FWIDTH=HWIDTH
        Q1=XK2*S0**.5
        V1=Q1/A2
        WRITE(NT,435)
435     FORMAT(/3X,'***STREETFLOW SPLITS OVER STREET-CROWN***')
        ISPLIT=1
        Q=Q-Q1
        TESTK=Q/S0**.5
        GOTO 300
437     CONTINUE
        WRITE(NT,438)CROWN,HWIDTH,V1,DN,FWIDTH,V
438     FORMAT(/5X,'FULL DEPTH(FEET) = ',F7.2,3X,'FLOODWIDTH(FEET) = ',F7.2,
     C  /,5X,'FULL HALF-STREET VELOCITY(FEET/SEC.) = ',F7.2,/,
     C  5X,'SPLIT DEPTH(FEET) = ',F7.2,3X,'SPLIT FLOODWIDTH(FEET) = ',F7.2,
     C  /,5X,'SPLIT VELOCITY(FEET/SEC.) = ',F7.2,/)
        V=V1
        DN=CROWN
        FWIDTH=HWIDTH
        GOTO 3000
440     CONTINUE
        WRITE(NT,441)
441     FORMAT(/3X,'***STREET FLOWING FULL***',/)
        DMAX=CROWN+5.
```

```
      DMIN=CROWN
      DO 450 I=1,12
      TEST=.5*(DMAX+DMIN)
      X=TEST-CROWN
      DN=CROWN+X
      A=HWIDTH*X+A2
      WP=WP2
      IF(DN.LE.CURB)WP=X+WP2
      HR=A/WP
      XK=1.486/XN*A*HR**.6667
      XZ=XK-TESTK
      XD=DN-DMAX
      IF(ABS(XD).LT..01)GOTO 3090
      IF(XZ)443,3090,445
443   DMIN=TEST
      GOTO 450
445   DMAX=TEST
450   CONTINUE
      ERROR=1.
      GO TO 3000
480   FWIDTH=HWIDTH
      V=Q/A
      GO TO 3000
2900  ERROR=2.
      WRITE(NT,6)
      WRITE(NT,2901)
2901  FORMAT(5X,'==> NO CONVERGENCE IN STREET FLOW "BELOW-CROWN" MODEL')
      WRITE(NT,6)
      GOTO 4000
3000  CONTINUE
      IF(ERROR.EQ.0.)GOTO 3100
      WRITE(NT,6)
      WRITE(NT,3005)
3005  FORMAT(5X,'==> MODELED FLOW DEPTH EXCEEDS 5-FEET ABOVE STREET CROWN')
      WRITE(NT,6)
      GOTO 4000
3090  FWIDTH=HWIDTH
      V=Q/A
      DV=V*DN
3100  WRITE(NT,180)
      WRITE(NT,3105)
3105  FORMAT(5X,'STREETFLOW MODEL RESULTS:')
      WRITE(NT,6)
      IF(DN.LE.CURB)GOTO 3109
      WRITE(NT,3107)
3107  FORMAT(5X,'NOTE: STREETFLOW EXCEEDS TOP OF CURB.',/,
     C 10X,'THE FOLLOWING STREETFLOW RESULTS ARE BASED ON THE ',
     . 'ASSUMPTION',
     C/,10X,'THAT NEGLIBLE FLOW OCCURS OUTSIDE OF THE STREET CHANNEL.',
     .  /,
     C 10X,'THAT IS, ALL FLOW ALONG THE PARKWAY, ETC., IS NEGLECTED.'/)
3109  CONTINUE
      DV=DN*V
      WRITE(NT,3110)DN,FWIDTH,V,DV
3110  FORMAT(5X,'STREET FLOWDEPTH(FEET) = ',F5.2,/,
     C 5X,'HALFSTREET FLOODWIDTH(FEET) = ',F7.2,/,
     C 5X,'AVERAGE FLOW VELOCITY(FEET/SEC.) = ',F7.2,/,
     C 5X,'PRODUCT OF DEPTH&VELOCITY = ',F7.2)
4000  CONTINUE
C
```

```
C  FORMATS
C
180      FORMAT(76('='))
187      FORMAT(76('*'))
6        FORMAT(76('-'))
C
         RETURN
         END
```

```
**********DESCRIPTION OF RESULTS*******************************************
*                                                                         *
*                 SYMMETRICAL STREETFLOW INFORMATION                      *
*                                                                         *
***************************************************************************

===========================================================================
     STREETFLOW MODEL INPUT INFORMATION:
---------------------------------------------------------------------------
     CONSTANT STREET GRADE(FEET/FEET) =  .005000
     CONSTANT STREET FLOW(CFS) =   25.00
     CONSTANT SYMMETRICAL STREET HALF-WIDTH(FEET) =   22.00
     CONSTANT SYMMETRICAL STREET CROSSFALL(DECIMAL) =  .020000
     AVERAGE STREETFLOW FRICTION FACTOR(MANNING) =  .015000
     CONSTANT SYMMETRICAL CURB HEIGTH(FEET) =    .67
     CONSTANT SYMMETRICAL GUTTER-WIDTH(FEET) =   2.00
     CONSTANT SYMMETRICAL GUTTER-HIKE(FEET) =   .16700
     CONSTANT SYMMETRICAL GUTTER-LIP(FEET) =   .03125
     FLOW ASSUMED TO FILL STREET ON ONE SIDE, AND THEN SPLITS
===========================================================================
   ***STREETFLOW SPLITS OVER STREET-CROWN***
---------------------------------------------------------------------------
     FULL DEPTH(FEET) =     .60   FLOODWIDTH(FEET) =   22.00
     FULL HALF-STREET VELOCITY(FEET/SEC.) =    2.72
     SPLIT DEPTH(FEET) =     .58   SPLIT FLOODWIDTH(FEET) =   21.06
     SPLIT VELOCITY(FEET/SEC.) =    2.45
---------------------------------------------------------------------------
===========================================================================
     STREETFLOW MODEL RESULTS:
---------------------------------------------------------------------------
     STREET FLOWDEPTH(FEET) =   .60
     HALFSTREET FLOODWIDTH(FEET) =   22.00
     AVERAGE FLOW VELOCITY(FEET/SEC.) =    2.72
     PRODUCT OF DEPTH&VELOCITY =    1.63
---------------------------------------------------------------------------
```

PROGRAM 15: DATA ENTRY

```
---DATA ENTRY FOR OPEN-CHANNEL PIPEFLOW JUNCTION ANALYSIS---PAGE 1



Enter assumed UPSTREAM flowdepth(FEET).............. ===>    "DEPTH1"
(NOTE: ENTER A [0] TO ASSUME NORMAL DEPTH)
:ALLOWABLE VALUES ARE [0] TO [100   ]


Enter assumed DOWNSTREAM flowdepth(FEET)............ ===>    "DEPTH2"
(NOTE: ENTER A [0] TO ASSUME NORMAL DEPTH)
:ALLOWABLE VALUES ARE [0] TO [100   ]







--------------------------------------------------------------------------------
    TYPE: EXIT to leave program ; TOP  to go to top of page
          MAIN to go to main menu
```

```
---DATA ENTRY FOR  ***UPSTREAM***  MAINLINE PIPE---PAGE 2

Enter pipeflow(CFS)................................. ===>    "QQ(1)"
:ALLOWABLE VALUES ARE [0] TO [10000 ]

Enter pipe diameter(INCHES)......................... ===>    "D(1)"
:ALLOWABLE VALUES ARE [3] TO [240   ]

Enter pipe friction factor(Manning)................. ===>    "XN(1)"
:ALLOWABLE VALUES ARE [.008   ] TO [.9999 ]

Enter pipe slope(DECIMAL)........................... ===>    "S(1)"
:ALLOWABLE VALUES ARE [.0001 ] TO [.25    ]

Enter pipe flowline elevation at junction(FEET)..... ===>    "FL(1)"
:ALLOWABLE VALUES ARE [-100   ] TO [10000 ]

Enter pipe angle(DEGREES) to downstream pipe........ ===>    "ANGLE(1)"
:ALLOWABLE VALUES ARE [0      ] TO [89    ]
*EQUATIONS MAY BECOME INVALID FOR ANGLES > 30 DEGREES

--------------------------------------------------------------------------------
    TYPE: EXIT to leave program ; TOP  to go to top of page
                                ; BACK to go back one page
```

```
---DATA ENTRY FOR ***LATERAL  1*** MAINLINE PIPE---PAGE 3

Enter pipeflow(CFS)................................. ===>     "QQ(3)"
:ALLOWABLE VALUES ARE [0] TO [10000 ]

Enter pipe diameter(INCHES)......................... ===>     "D(3)"
:ALLOWABLE VALUES ARE [3] TO [240   ]

Enter pipe friction factor(Manning)................. ===>     "XN(3)"
:ALLOWABLE VALUES ARE [.008   ] TO [.9999 ]

Enter pipe slope(DECIMAL)........................... ===>     "S(3)"
:ALLOWABLE VALUES ARE [.0001 ] TO [.25    ]

Enter pipe flowline elevation at junction(FEET)..... ===>     "FL(3)"
:ALLOWABLE VALUES ARE [-100   ] TO [10000 ]

Enter pipe angle(DEGREES) to downstream pipe........ ===>     "ANGLE(3)"
:ALLOWABLE VALUES ARE [1      ] TO [90    ]


--------------------------------------------------------------------------------
    TYPE: EXIT to leave program ; TOP  to go to top of page
                                ; BACK to go back one page
```

```
---DATA ENTRY FOR ***LATERAL  2*** MAINLINE PIPE---PAGE 4

Enter pipeflow(CFS)................................. ===>     "QQ(4)"
:ALLOWABLE VALUES ARE [0] TO [10000 ]

Enter pipe diameter(INCHES)......................... ===>     "D(4)"
:ALLOWABLE VALUES ARE [3] TO [240   ]

Enter pipe friction factor(Manning)................. ===>     "XN(4)"
:ALLOWABLE VALUES ARE [.008   ] TO [.9999 ]

Enter pipe slope(DECIMAL)........................... ===>     "S(4)"
:ALLOWABLE VALUES ARE [.0001 ] TO [.25    ]

Enter pipe flowline elevation at junction(FEET)..... ===>     "FL(4)"
:ALLOWABLE VALUES ARE [-100   ] TO [10000 ]

Enter pipe angle(DEGREES) to downstream pipe........ ===>     "ANGLE(4)"
:ALLOWABLE VALUES ARE [1      ] TO [90    ]


--------------------------------------------------------------------------------
    TYPE: EXIT to leave program ; TOP  to go to top of page
                                ; BACK to go back one page
```

```
---DATA ENTRY FOR ***DOWNSTREAM*** MAINLINE PIPE---PAGE 5

DOWNSTREAM PIPEFLOW EQUALS THE SUM OF THE UPSTREAM AND
LATERAL FLOWS = [    30.00] + [    20.00] + [    10.00] = [       60.00]

Enter pipe diameter(INCHES).......................... ===>    "D(2)"
:ALLOWABLE VALUES ARE [3] TO [240   ]

Enter pipe friction factor(Manning)................. ===>    "XN(2)"
:ALLOWABLE VALUES ARE [.008   ] TO [.9999 ]

Enter pipe slope(DECIMAL)........................... ===>    "S(2)"
:ALLOWABLE VALUES ARE [.0001 ] TO [.25   ]

Enter pipe flowline elevation at junction(FEET)..... ===>    "FL(2)"
:ALLOWABLE VALUES ARE [-100  ] TO [10000 ]




-------------------------------------------------------------------------------
    TYPE: EXIT to leave program ; TOP  to go to top of page
                                ; BACK to go back one page
```

PROGRAM 15

```
C  ------------------------------------------------------------------
         SUBROUTINE OPCHAN
C  ------------------------------------------------------------------
C
C   ANALYSIS OF PIPEFLOW OPEN CHANNEL
C
         DIMENSION QQ(4),D(4),XN(4),ANGLE(4),S(4),FL(4)
         COMMON /NUT/NUT
C
C
         NT=NUT
C
C   INITIALIZE ARRAYS
         DO 2 I=1,4
         QQ(I)=0.
         D(I)=0.
         FL(I)=0.
         XN(I)=0.
         ANGLE(I)=0.
         S(I)=0.
2        CONTINUE
C
C..READ DATA INPUT
         READ FREE(5)DEPTH1,DEPTH2
         DO 5 I=1,4
         READ FREE(5)QQ(I),D(I),XN(I),S(I),FL(I),ANGLE(I)
5        CONTINUE
         QQ(2)=QQ(1)+QQ(3)+QQ(4)
C
C
         DELZ=FL(1)-FL(2)
         CALL OPNJ1(QQ,D,XN,ANGLE,S,FL,DEPTH1,DEPTH2,DELZ,NT)
C
         RETURN
         END
C
C
C  ------------------------------------------------------------------
         SUBROUTINE OPNJ1(QQ,D,XN,ANGLE,S,FL,DEPTH1,DEPTH2,DELZ,NT)
C  ------------------------------------------------------------------
C
C        COMPUTATION/LOGIC ROUTINE
C
         DIMENSION QQ(4),D(4),XN(4),SO(4),YYC(4),YYN(4),ANGLE(4),R(4)
         DIMENSION S(4),FL(4)
C
         DELTA(YY,RR)=ACOS((RR-YY)/RR)
         AREA(YY,RR)=RR*RR*(ANG-.5*SIN(2.*ANG))
         TW(YY,RR)=2.*RR*SIN(ANG)
         YCRIT(YY,Q,RR)=1.-Q*Q*TW(YY,RR)/(G*AREA(YY,RR)**3.)
         PER(RR)=2.*RR*ANG
         YNORM(YY,Q,RR)=1.-Q*Q*CON/(AREA(YY,RR)**3.3333/PER(RR)**1.3333)
C
C        INITIALIZE ARRAYS
C
         DO 900 J=1,4
         YYC(J)=0.
900      YYN(J)=0.
         DEPTH3=0.
```

```
          DEPTH4=0.
C   DISPLAY INPUT INFORMATION
          WRITE(NT,187)
          WRITE(NT,51)
51        FORMAT(1X,'>>>>PIPE-FLOW JUNCTION INPUT INFORMATION<<<<')
          WRITE(NT,6)
          WRITE(NT,42)
42       FORMAT(8X,'PIPE',7X,'FLOW',4X,'DIAMETER',4X,'SLOPE',3X,'FRICTION'
     C   ,4X,'ANGLE',4X,'FLOWLINE',
     C   /,19X,'(CFS)',3X,'(INCHES)',2X,'(DECIMAL)',2X,'FACTOR',
     C   3X,'(DEGREES)',1X,'ELEVATION')
          WRITE(NT,43)(QQ(J),D(J),S(J),XN(J),ANGLE(J),FL(J),J=1,4)
43        FORMAT(6X,'UPSTREAM ',F10.2,F10.3,F10.5,F10.4,F10.3,F10.2/,
     C   5X,'DOWNSTREAM',F10.2,F10.3,F10.5,F10.4,F10.3,F10.2,/,
     C   5X,'LATERAL #1',F10.2,F10.3,F10.5,F10.4,F10.3,F10.2,/,
     C   5X,'LATERAL #2',F10.2,F10.3,F10.5,F10.4,F10.3,F10.2,/)
          WRITE(NT,44)DEPTH1,DEPTH2
44        FORMAT(5X,'MAINLINE FLOWDEPTH INPUT INFORMATION:',/,
     C   5X,'UPSTREAM PIPEFLOW DEPTH(FEET):',F10.2,/,
     C   5X,'DOWNSTREAM PIPEFLOW DEPTH(FEET):',F10.2)
C
C   CHECK FOR PROPER LOWER LIMIT OF INPUT VALUES
          DO 1050 I=1,4
          IF(D(I).EQ.0.)D(I)=.0003
          IF(FL(I).EQ.0.)FL(I)=.000001
          IF(XN(I).EQ.0.)XN(I)=.000001
          IF(ANGLE(I).EQ.0.)ANGLE(I)=.000001
C-------DETERMINE YC,YN FOR POINTS 1,2(1,2=UPSTREAM,DOWNSTREAM)
90        ICOUNT=0
          KK=0
          GAM=62.4
          G=32.2
100       ICOUNT=ICOUNT+1
          IF(ICOUNT.EQ.5)GOTO 200
          J=ICOUNT
          YC=0.
          IF(QQ(J) .LT. .001)GO TO 121
          DIAM=D(J)/12.
          R(J)=DIAM/2.
          CON=(XN(J)/1.486)**2./S(J)
          DMAX=.82*DIAM
112       UP=DIAM
          DN=0.
          YC=DMAX/2.
          DO 120 I=1,22
          ANG=DELTA(YC,R(J))
          IF(YCRIT(YC,QQ(J),R(J)))114,121,115
114       DN=YC
          GOTO 120
115       UP=YC
120       YC=(UP+DN)*.5
121       YYC(J)=YC
          YN=0.
          IF(QQ(J).LT. .001)GO TO 131
          TEST=.463*DIAM**2.6667*S(J)**.5/XN(J)
          IF(QQ(J).LT.TEST) GOTO 122
          YN=DIAM
          IF(J.EQ.2.AND.DEPTH2.EQ.0.)KODE=1
          GOTO 131
122       UP=DMAX
```

```
        DN=0.
        YN=DMAX/2.
        DO 130 I=1,22
        ANG=DELTA(YN,R(J))
        IF(YNORM(YN,QQ(J),R(J)))123,131,124
123     DN=YN
        GOTO 130
124     UP=YN
130     YN=(UP+DN)*.5
131     YYN(J)=YN
        GOTO 100
C  DISPLAY NORMAL AND CRITICAL DEPTH INFORMATION
200     WRITE(NT,5)
        WRITE(NT,201)
201     FORMAT(5X,'PIPEFLOW NORMAL AND CRITICAL DEPTH INFORMATION:')
        WRITE(NT,6)
        WRITE(NT,202)
202     FORMAT(8X,'PIPE',4X,'CRITICAL DEPTH',2X,'NORMAL DEPTH',/,
     C  20X,'(FEET)',9X,'(FEET)')
        WRITE(NT,203)(YYC(J),YYN(J),J=1,4)
203     FORMAT(6X,'UPSTREAM',6X,F6.3,9X,F6.3,/,
     C  5X,'DOWNSTREAM',5X,F6.3,9X,F6.3,/,5X,'LATERAL #1',5X,F6.3,
     C  9X,F6.3,/,5X,'LATERAL #2',5X,F6.3,9X,F6.3)
C
C       ASSIGN DEFAULT CONTROL VALUES
        IF(DEPTH1.EQ.0.)DEPTH1=YYN(1)
        IF(DEPTH2.EQ.0.)DEPTH2=YYN(2)
        IF(DEPTH3.EQ.0.)DEPTH3=YYN(3)
        IF(DEPTH4.EQ.0.)DEPTH4=YYN(4)
C
C  DISPLAY FORMULA
C
        WRITE(NT,990)
990    FORMAT(/5X,'PRESSURE-PLUS-MOMENTUM DETERMINATION BASED ON VARIA',
     C  'BLE:',/,5X,'"BALANCE" = (Z+D1-D2)*(A1+A2)*G/2.-Q2*Q2/A2'
     C  '+Q1*Q1*COS(ANGLE1)/A1',/,5X,'+Q3*Q3*COS(ANGLE3)/A3'
     C  '+Q4*Q4*COS(ANGLE4)/A4')
        IF(KODE.EQ.0)GOTO 7
        WRITE(NT,6)
        WRITE(NT,8)
8       FORMAT(5X,'POSSIBLE LOGIC ERROR: DOWNSTREAM PIPEFLOW MAY',/,
     .  5X,'BE UNDER PRESSURE.  PROGRAM USER MUST THEREFORE SPECIFY',/,
     .  5X,'A PROPER "CONTROL DEPTH" FOR ASSUMED DOWNSTREAM CONTROL',/,
     .  5X,'AND NOT USE THE PROGRAM-COMPUTED NORMAL DEPTH FOR THE',/,
     .  5X,'ASSUMED DOWNSTREAM CONTROL.')
        GOTO 3000
C
C
C-------DETERMINE CONTROLS:
C
C       CHECK FOR PRESSURE FLOW
7       DMAX=.82*D(1)/12.
        DMAXT=D(1)/12.
        IF(DEPTH1.LE.DMAXT)GOTO 210
        WRITE(NT,6)
5       FORMAT(76('='))
6       FORMAT(76('-'))
        WRITE(NT,205)
205     FORMAT(5X,'PRESSURE FLOW ASSUMED THROUGH JUNCTION DUE TO',/,
        5X,'USER-SPECIFIED PRESSURE FLOW ASSUMPTION FOR THE',/,
```

```
     .  5X,'UPSTREAM PIPEFLOW.  SUGGEST USER RE-EXAMINE ASSUMED',/,
     .  5X,'PIPEFLOW FLOWDEPTHS FOR POSSIBLE CRITICAL DEPTH OR',/,
     .  5X,'SOFFIT CONTROL IN UPSTREAM PIPE, OR PRESSURE',/,
     .  5X,'FLOW FOR BOTH DOWNSTREAM AD UPSTREAM PIPEFLOW.')
        GOTO 3000
C  OPEN-CHANNEL FLOW
C
C       DOWNSTREAM FLOW IS SUBCRITICAL OR CONTROL EXCEEDS CRITICAL DEPTH:
C
C       [KONTROL=1]=> DOWNSTREAM CONTROL; LET DEPTH2=DEPTH2,FOR DEPTH2>YYC(2)
210     CONTINUE
        IF(DEPTH2.LT.YYC(2))GOTO 250
        KONTROL=1
        IF(DEPTH1.LT.YYC(1))GOTO 230
C.......CHECK FOR WASHOUT DUE TO LATERALS OR DELZ-DROP
        WRITE(NT,6)
        WRITE(NT,991)
991     FORMAT(5X,'CHECK FOR JUNCTION WASHOUT DUE TO LATERALS OR',
     C  ' JUNCTION DROP:')
        WRITE(NT,979)
        WRITE(NT,980)
        CALL FPM(DELZ,DEPTH1,DEPTH2,R,QQ,YYC,YYN,ANGLE,FL,TEST,NT)
        IF(TEST.LT. .001)GO TO 992
        WRITE(NT,6)
        WRITE(NT,993)
993     FORMAT(5X,'*JUNCTION DROP IN ELEVATION OR LATERAL PRESSURE-',
     C  'PLUS-MOMENTUM',/,5X,'*CAUSES UPSTREAM FLOWS TO DOMINATE ',
     C'HYDRAULICS.',/,5X,'*SUGGEST REANALYZE JUNCTION FOR HYDRAULIC ',
     C  'CONTROL.')
        GOTO 3000
992     WRITE(NT,971)
971     FORMAT(5X,'*DOWNSTREAM PIPEFLOW DEPTH IS ASSUMED AS HYDRAULIC ',
     C  'CONTROL')
        GOTO 300
C  CHECK FOR JUNCTION WASHOUT
230     WRITE(NT,6)
        WRITE(NT,979)
        WRITE(NT,980)
        CALL FPM(DELZ,DEPTH1,DEPTH2,R,QQ,YYC,YYN,ANGLE,FL,TEST,NT)
        WRITE(NT,972)
972     FORMAT(5X,'UPSTREAM PIPEFLOW IS SUPERCRITICAL, AND DOWNSTREAM',
     C  /,5X,'PIPEFLOW IS SUBCRITICAL OR UNDER PRESSURE:')
        IF(TEST.LT.0.)WRITE(NT,973)
973     FORMAT(5X,'*DOWNSTREAM FLOW DOMINATES JUNCTION HYDRAULICS',
     C  /,5X,'*HYDRAULIC JUMP MUST OCCUR UPSTREAM OF JUNCTION.')
        IF(TEST.LT.0.)GOTO 300
        WRITE(NT,974)
974     FORMAT(5X,'*UPSTREAM FLOW DOMINATES JUNCTION HYDRAULICS')
C       [KONTROL=5]=> UPSTREAM FLOW IS SUPERCRITICAL AND WASHOUT OCCURS
        KONTROL=5
        GOTO 300
C
C  DOWNSTREAM FLOW IS SUPERCRITICAL:
C
C       [KONTROL=2]=> UPSTREAM FLOW IS SUBCRITICAL;
C                         LET DEPTH2=YYC(2)
250     IF(DEPTH1.LE.YYC(1))GOTO 275
        IF(DEPTH2.LT.YYC(2))DEPTH2=YYC(2)
        WRITE(NT,975)
975     FORMAT(/5X,'DOWNSTREAM FLOW IS SUPERCRITICAL AND UPSTREAM FLOW'
```

```
     C  ,/,5X,'IS SUBCRITICAL: CRITICAL DEPTH IS USED AS A DOWNSTREAM
     C  ,'CONTROL')
        KONTROL=2
        GOTO 300
C
C       [KONTROL=3]=> UPSTREAM FLOW IS SUPERCRITICAL AND JUMP FAILS
C       [KONTROL=4]=> UPSTREAM FLOW IS SUPERCRITICAL AND JUMP OCCURS
C=====>>TEST FOR HYDRAULIC JUMP
275     CONTINUE
        WRITE(NT,6)
        WRITE(NT,976)
        WRITE(NT,979)
979     FORMAT(5X,'PIPEFLOW FORCE-PLUS-MOMENTUM DETERMINATION(NEGLECT
     C  MINOR LOSSES)')
        WRITE(NT,980)
980     FORMAT(12X,'UPSTREAM DOWNSTREAM LATERAL#1 LATERAL#2  BALANCE',/,
     C  12X,'DEPTH(FT) DEPTH(FT) DEPTH(FT) DEPTH(FT)   (FT**4)')
        CALL FPM(DELZ,DEPTH1,YYC(2),R,QQ,YYC,YYN,ANGLE,FL,TEST,NT)
976     FORMAT(5X,'UPSTREAM FLOW IS SUPERCRITICAL; CHECK FOR HYDRAULIC',
     C  ' JUMP:')
        IF(TEST.LE.0.)GOTO 290
C NO HYDRAULIC JUMP
        WRITE(NT,977)
977     FORMAT(5X,'*UPSTREAM FLOW DOMINATES JUNCTION HYDRAULICS:',/,
     C  5X,'*NO HYDRAULIC JUMP OCCURS AT JUNCTION.')
        KONTROL=3
        GOTO 300
C  HYDRAULIC JUMP OCCURS UPSTREAM OF JUNCTION
290     KONTROL=4
        WRITE(NT,978)
978     FORMAT(5X,'*HYDRAULIC JUMP OCCURS UPSTREAM OF JUNCTION:',/,
     C  5X,'*CRITICAL DEPTH IS ASSUMED AS A DOWNSTREAM HYDRAULIC ',
     C  'CONTROL.')
        DEPTH2=YYC(2)
300     CONTINUE
C
C       ITERATION
C
        GOTO(400,400,410,400,410)KONTROL
C====>>>DOWNSTREAM CONTROL
C
C:      CHECK IF JUNCTION SEALS:
400     DMAX=2.*R(1)
        SIGN=1.
        WRITE(NT,6)
        WRITE(NT,981)
981     FORMAT(5X,'CHECK IF JUNCTION SEALS DUE TO DOWNSTREAM CONTROL:')
        WRITE(NT,979)
        WRITE(NT,980)
        CALL FPM(DELZ,DMAX,DEPTH2,R,QQ,YYC,YYN,ANGLE,FL,TEST,NT)
        IF(TEST.GT.0.)WRITE(NT,982)
982     FORMAT(5X,'*UPSTREAM FLOW ASSUMED NOT SEALED.')
        IF(TEST.GT.0.)GOTO 420
C.......JUNCTION IS ASSUMED SEALED FOR D1>DIAM1
        WRITE(NT,6)
        WRITE(NT,401)
401     FORMAT(5X,'*UPSTREAM WATER DEPTH EXCEEDS PIPE DIAMETER:',/,
     C  5X,'*SUGGEST REANALYZE JUNCTION AS UNDER PRESSURE-FLOW ',
     C  'CONDITIONS.')
        GOTO 3000
```

```
C====>>>UPSTREAM CONTROL
C
410     SIGN=-1.
        DTOP=YYC(2)
        DLOW=0.
        GOTO 450
C  BEGIN DEPTH-DETERMINATION
C
C:      DOWNSTREAM CONTROL:
420     DTOP=2.*R(1)*.995
        DLOW=YYC(1)
450     DG=.5*(DTOP+DLOW)
C
        WRITE(NT,5)
        WRITE(NT,979)
        WRITE(NT,980)
        WRITE(NT,6)
        IF(SIGN.LT.0.)GOTO 500
C:      DOWNSTREAM CONTROL:
        DO 475 J=1,12
        CALL FPM(DELZ,DG,DEPTH2,R,QQ,YYC,YYN,ANGLE,FL,TEST,NT)
        IF(TEST)460,2000,470
460     DLOW=DG
        GOTO 475
470     DTOP=DG
475     DG=.5*(DLOW+DTOP)
        GOTO 2000
C:      UPSTREAM CONTROL:
500     DO 525 J=1,13
        CALL FPM(DELZ,DEPTH1,DG,R,QQ,YYC,YYN,ANGLE,FL,TEST,NT)
        IF(TEST)510,2000,520
510     DLOW=DG
        GOTO 525
520     DTOP=DG
525     DG=.5*(DLOW+DTOP)
2000    CONTINUE
C
C       OUTPUT RESULTS
C
        WRITE(NT,6)
        IF(KONTROL.LT.3)WRITE(NT,2101)
        IF(KONTROL.EQ.3)WRITE(NT,2102)
        IF(KONTROL.EQ.4)WRITE(NT,2101)
        IF(KONTROL.EQ.5)WRITE(NT,2102)
2101    FORMAT(5X,'DOWNSTREAM CONTROL ASSUMED AT JUNCTION')
2102    FORMAT(5X,'UPSTREAM CONTROL ASSUMED AT JUNCTION')
        WRITE(NT,6)
        IF(KONTROL.EQ.3.OR.KONTROL.EQ.5)WRITE(NT,2100)DEPTH1,DG
        IF(KONTROL.NE.3.AND.KONTROL.NE.5)WRITE(NT,2100)DG,DEPTH2
2100    FORMAT(5X,'COMPUTED UPSTREAM PIPEFLOW DEPTH(FEET) = ',F6.3,/,
     C  5X,'COMPUTED DOWNSTREAM PIPEFLOW DEPTH(FEET) = ',F6.3)
C
3000    CONTINUE
C
187     FORMAT(76('*'))
C
        RETURN
        END
C
C
```

```
C  ----------------------------------------------------------------
         SUBROUTINE FPM(Z,D1,D2,R,Q,YYC,YYN,ANGLE,FL,TEST,NT)
C  ----------------------------------------------------------------
C  SUBROUTINE DETERMINES:   [TEST] = {FPM(IN)-FPM(OUT)}
C
         DIMENSION R(4),Q(4),YYC(4),YYN(4),ANGLE(4),FL(4)
C
C
         DELTA(YY,RR)=ACOS((RR-YY)/RR)
         AREA(YY,RR)=RR*RR*(ANG-.5*SIN(2.*ANG))
         A3=1.
         A4=1.
         D3=0.
         D4=0.
         ANG=DELTA(D1,R(1))
         F=3.1415926/180.
         A1=AREA(D1,R(1))
         A2=R(2)*R(2)*3.141593
         X1=2.*R(2)
         IF(D2.GE.X1)GOTO 3
         ANG=DELTA(D2,R(2))
         A2=AREA(D2,R(2))
3        T=.5*(D1+D2+FL(1)+FL(2))
         IF(Q(3).LT. .001)GOTO 20
C-------LATERAL: LINE#3
         ANG=DELTA(YYN(3),R(3))
         A3=AREA(YYN(3),R(3))
         D3=YYN(3)
         IF(YYC(3).GE.YYN(3))GOTO 20
C:       MILD FLOW
         X=T-FL(3)
         IF(YYC(3).GE.X)GOTO 5
         A3=R(3)*R(3)*3.141593
         X1=R(3)*2.
         IF(X.LT.X1)GOTO 10
         D3=X1
         IF(X.GT.X1)D3=X
         IF(D3.GE.X1)GOTO 20
5        ANG=DELTA(YYC(3),R(3))
         A3=AREA(YYC(3),R(3))
         D3=YYC(3)
         GOTO 20
10       ANG=DELTA(X,R(3))
         A3=AREA(X,R(3))
         D3=X
C-------LATERAL: LINE#4
20       IF(Q(4).LT. .001)GOTO 50
         ANG=DELTA(YYN(4),R(4))
         A4=AREA(YYN(4),R(4))
         D4=YYN(4)
         IF(YYC(4).GE.YYN(4))GOTO 40
C:       MILD FLOW
         X=T-FL(4)
         IF(YYC(4).GE.X)GOTO 25
         A4=R(4)*R(4)*3.141593
         X1=R(4)*2.
         IF(X.LT.X1)GOTO 30
         D4=X1
         IF(X.GT.X1)D4=X
         IF(D4.GE.X1)GOTO 40
```

```
25      ANG=DELTA(YYC(4),R(4))
        A4=AREA(YYC(4),R(4))
        D4=YYC(4)
        GOTO 40
30      ANG=DELTA(X,R(4))
        A4=AREA(X,R(4))
        D4=X
40      CONTINUE
50      TEST=(Z+D1-D2)*(A1+A2)*16.1-Q(2)*Q(2)/A2+Q(1)*Q(1)*COS(ANGLE(1)*
     C  F)/A1+Q(3)*Q(3)*COS(ANGLE(3)*F)/A3+Q(4)*Q(4)*COS(ANGLE(4)*F)/A4
        WRITE(NT,100)D1,D2,D3,D4,TEST
100     FORMAT(10X,4F10.3,F11.0)
C
        RETURN
        END
```

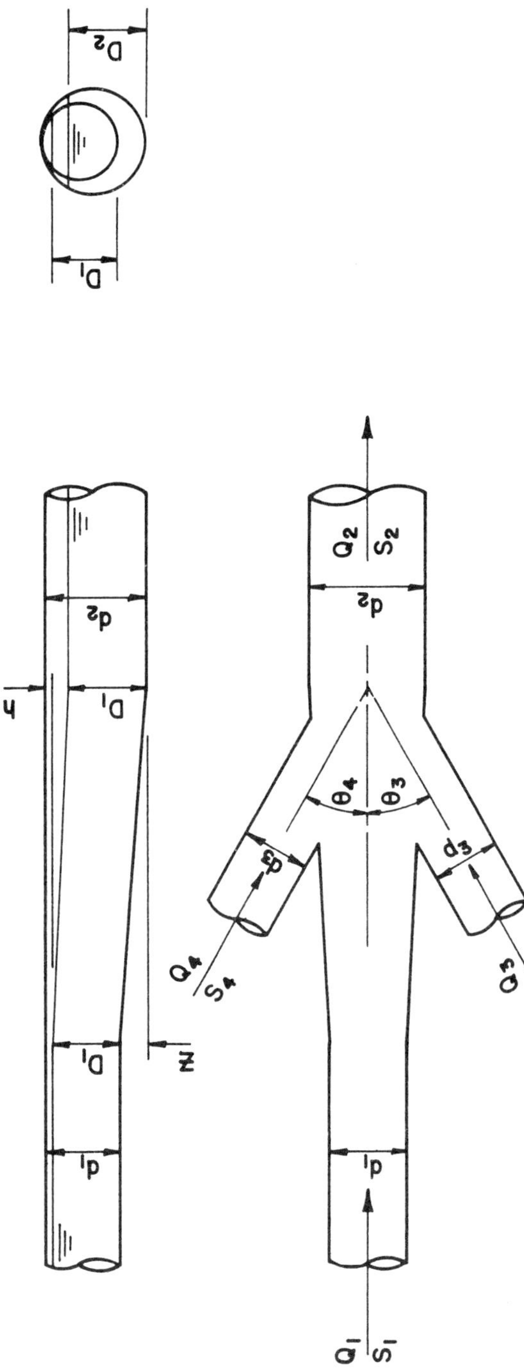

EXAMPLE PROBLEM

```
==========================================================================
                  HYDRAULIC ELEMENTS - I  PROGRAM PACKAGE
==========================================================================


**********DESCRIPTION OF RESULTS********************************************
*                          EXAMPLE PROBLEM                                *
*                                                                         *
*                                                                         *
****************************************************************************

****************************************************************************
 >>>>PIPE-FLOW JUNCTION INPUT INFORMATION<<<<
--------------------------------------------------------------------------
        PIPE        FLOW    DIAMETER     SLOPE    FRICTION     ANGLE    FLOWLINE
                    (CFS)   (INCHES)   (DECIMAL)   FACTOR    (DEGREES) ELEVATION
      UPSTREAM     100.00    36.000     .01000     .0130        .000     101.50
     DOWNSTREAM    125.00    54.000     .01000     .0130        .000     100.00
     LATERAL #1     25.00    30.000     .01000     .0130      45.000     100.50
     LATERAL #2       .00      .000     .00000     .0000        .000        .00

     MAINLINE FLOWDEPTH INPUT INFORMATION:
     UPSTREAM PIPEFLOW DEPTH(FEET):        .00
     DOWNSTREAM PIPEFLOW DEPTH(FEET):        .00
==========================================================================
     PIPEFLOW NORMAL AND CRITICAL DEPTH INFORMATION:
--------------------------------------------------------------------------
        PIPE    CRITICAL DEPTH  NORMAL DEPTH
                    (FEET)         (FEET)
      UPSTREAM      2.895          3.000
     DOWNSTREAM     3.291          2.605
     LATERAL #1     1.703          1.409
     LATERAL #2      .000           .000

     PRESSURE-PLUS-MOMENTUM DETERMINATION BASED ON VARIABLE:
     "BALANCE" = (Z+D1-D2)*(A1+A2)*G/2.-Q2*Q2/A2+Q1*Q1*COS(ANGLE1)/A1
     +Q3*Q3*COS(ANGLE3)/A3+Q4*Q4*COS(ANGLE4)/A4

     DOWNSTREAM FLOW IS SUPERCRITICAL AND UPSTREAM FLOW
     IS SUBCRITICAL: CRITICAL DEPTH IS USED AS A DOWNSTREAM CONTROL
--------------------------------------------------------------------------
     CHECK IF JUNCTION SEALS DUE TO DOWNSTREAM CONTROL:
     PIPEFLOW FORCE-PLUS-MOMENTUM DETERMINATION(NEGLECT MINOR LOSSES)
            UPSTREAM DOWNSTREAM LATERAL#1 LATERAL#2  BALANCE
            DEPTH(FT) DEPTH(FT) DEPTH(FT) DEPTH(FT)   (FT**4)
               3.000     3.291     1.409      .000       696.
     *UPSTREAM FLOW ASSUMED NOT SEALED.
==========================================================================
     PIPEFLOW FORCE-PLUS-MOMENTUM DETERMINATION(NEGLECT MINOR LOSSES)
            UPSTREAM DOWNSTREAM LATERAL#1 LATERAL#2  BALANCE
            DEPTH(FT) DEPTH(FT) DEPTH(FT) DEPTH(FT)   (FT**4)
--------------------------------------------------------------------------
               2.940     3.291     1.409      .000       684.
               2.918     3.291     1.409      .000       680.
               2.907     3.291     1.409      .000       679.
               2.901     3.291     1.409      .000       678.
               2.898     3.291     1.409      .000       678.
               2.897     3.291     1.409      .000       678.
               2.896     3.291     1.409      .000       678.
```

```
             2.896     3.291     1.409      .000      678.
             2.896     3.291     1.409      .000      678.
             2.896     3.291     1.409      .000      678.
             2.896     3.291     1.409      .000      678.
             2.895     3.291     1.409      .000      678.
---------------------------------------------------------------------------
     DOWNSTREAM CONTROL ASSUMED AT JUNCTION
---------------------------------------------------------------------------
     COMPUTED UPSTREAM PIPEFLOW DEPTH(FEET) =  2.895
     COMPUTED DOWNSTREAM PIPEFLOW DEPTH(FEET) =  3.291
```

```
============================================================================
                  HYDRAULIC ELEMENTS - I  PROGRAM PACKAGE
============================================================================


**********DESCRIPTION OF RESULTS********************************************
*                         EXAMPLE PROBLEM                                  *
*                                                                          *
*                                                                          *
****************************************************************************

****************************************************************************
 >>>>PIPE-FLOW JUNCTION INPUT INFORMATION<<<<
----------------------------------------------------------------------------
        PIPE         FLOW    DIAMETER    SLOPE    FRICTION    ANGLE    FLOWLINE
                    (CFS)   (INCHES)  (DECIMAL)   FACTOR   (DEGREES) ELEVATION
      UPSTREAM     100.00    54.000    .02000      .0130       .000    100.00
     DOWNSTREAM    115.00    54.000    .02000      .0130       .000    100.00
     LATERAL #1     10.00    18.000    .00500      .0130     45.000    102.00
     LATERAL #2      5.00    18.000    .00500      .0130     45.000    102.00

     MAINLINE FLOWDEPTH INPUT INFORMATION:
     UPSTREAM PIPEFLOW DEPTH(FEET):          .00
     DOWNSTREAM PIPEFLOW DEPTH(FEET):          .00
============================================================================
     PIPEFLOW NORMAL AND CRITICAL DEPTH INFORMATION:
----------------------------------------------------------------------------
        PIPE     CRITICAL DEPTH  NORMAL DEPTH
                     (FEET)         (FEET)
      UPSTREAM        2.938          1.865
     DOWNSTREAM       3.156          2.017
     LATERAL #1       1.219          1.500
     LATERAL #2        .860           .901

     PRESSURE-PLUS-MOMENTUM DETERMINATION BASED ON VARIABLE:
     "BALANCE" = (Z+D1-D2)*(A1+A2)*G/2.-Q2*Q2/A2+Q1*Q1*COS(ANGLE1)/A1
     +Q3*Q3*COS(ANGLE3)/A3+Q4*Q4*COS(ANGLE4)/A4
----------------------------------------------------------------------------
     UPSTREAM FLOW IS SUPERCRITICAL; CHECK FOR HYDRAULIC JUMP:
     PIPEFLOW FORCE-PLUS-MOMENTUM DETERMINATION(NEGLECT MINOR LOSSES)
            UPSTREAM DOWNSTREAM LATERAL#1 LATERAL#2  BALANCE
            DEPTH(FT) DEPTH(FT) DEPTH(FT) DEPTH(FT)  (FT**4)
               1.865     3.156     1.219      .860       181.
     *UPSTREAM FLOW DOMINATES JUNCTION HYDRAULICS:
     *NO HYDRAULIC JUMP OCCURS AT JUNCTION.
============================================================================
     PIPEFLOW FORCE-PLUS-MOMENTUM DETERMINATION(NEGLECT MINOR LOSSES)
            UPSTREAM DOWNSTREAM LATERAL#1 LATERAL#2  BALANCE
            DEPTH(FT) DEPTH(FT) DEPTH(FT) DEPTH(FT)  (FT**4)
----------------------------------------------------------------------------
               1.865     1.578     1.219      .860      -939.
               1.865     2.367     1.219      .860       -11.
               1.865     2.761     1.219      .860       138.
               1.865     2.564     1.219      .860        80.
               1.865     2.466     1.219      .860        39.
               1.865     2.416     1.219      .860        15.
               1.865     2.392     1.219      .860         3.
               1.865     2.379     1.219      .860        -4.
               1.865     2.385     1.219      .860        -1.
```

```
              1.865     2.388     1.219     .860      1.
              1.865     2.387     1.219     .860      0.
              1.865     2.386     1.219     .860      0.
              1.865     2.387     1.219     .860      0.
---------------------------------------------------------------------------
     UPSTREAM CONTROL ASSUMED AT JUNCTION
---------------------------------------------------------------------------
     COMPUTED UPSTREAM PIPEFLOW DEPTH(FEET) =  1.865
     COMPUTED DOWNSTREAM PIPEFLOW DEPTH(FEET) =  2.386
===========================================================================
```

PROGRAM 16: DATA ENTRY

```
---DATA ENTRY FOR GRADUALLY VARIED FLOW WATER SURFACE DETERMINATION
   FOR CONSTANT SLOPE RECT/TRAP/V-SHAPED OR PIPE---PAGE 1

Enter constant channel slope(FEET/FEET)............. ===>    "SO"
:ALLOWABLE VALUES ARE [.00001] TO [.99    ]

Enter length(FEET) of channel with constant slope... ===>    "XL"
:ALLOWABLE VALUES ARE [1] TO [100000]

Enter constant channel flow(CFS).................... ===>    "Q"
:ALLOWABLE VALUES ARE [.01    ] TO [1000000    ]

Enter channel friction factor(Mannings)............. ===>    "RN"
:ALLOWABLE VALUES ARE [.008   ] TO [.9999 ]

Enter channel control-depth(FEET)................... ===>    "YCONT"
(NOTE: IF [0] IS ENTERED, CRITICAL DEPTH IS ASSUMED
 AS CONTROL
:ALLOWABLE VALUES ARE [0] TO [1000   ]

--------------------------------------------------------------------------------
TYPE: EXIT to leave program ; TOP  to go to top of page
      MAIN to go to main menu
```

```
---DATA ENTRY FOR GRADUALLY VARIED FLOW WATER SURFACE DETERMINATION
   FOR CONSTANT SLOPE RECTANGULAR CHANNEL---PAGE 2

Enter maximum number of intervals to be
used in profile..................................... ===>    "NN"
:ALLOWABLE VALUES ARE [10     ] TO [1000  ]

Enter channel basewidth(FEET)....................... ===>    "B"
:ALLOWABLE VALUES ARE [0] TO [1000   ]

--------------------------------------------------------------------------------
TYPE: EXIT to leave program ; TOP  to go to top of page
                            ; BACK to go back one page
```

```
---DATA ENTRY FOR GRADUALLY VARIED FLOW WATER SURFACE DETERMINATION
   FOR CONSTANT SLOPE TRAPEZOIDAL CHANNEL---PAGE 2

Enter maximum number of intervals to be
used in profile..................................... ===>    "NN"
:ALLOWABLE VALUES ARE [10    ] TO [1000  ]

Enter channel basewidth(FEET)....................... ===>    "B"
:ALLOWABLE VALUES ARE [0] TO [1000  ]

Enter channel Z value............................... ===>    "Z"
(NOTE: Z = QUOTIENT OF (HORIZONTAL)/(VERTICAL))
:ALLOWABLE VALUES ARE [0] TO [100   ]

--------------------------------------------------------------------------------
   TYPE: EXIT to leave program ; TOP  to go to top of page
                               ; BACK to go back one page
```

```
---DATA ENTRY FOR GRADUALLY VARIED FLOW WATER SURFACE DETERMINATION
   FOR CONSTANT SLOPE (V) -SHAPED CHANNEL---PAGE 2

Enter maximum number of intervals to be
used in profile..................................... ===>    "NN"
:ALLOWABLE VALUES ARE [10    ] TO [1000  ]



Enter channel Z value............................... ===>    "Z"
(NOTE: Z = QUOTIENT OF (HORIZONTAL)/(VERTICAL))
:ALLOWABLE VALUES ARE [0] TO [100   ]

--------------------------------------------------------------------------------
   TYPE: EXIT to leave program ; TOP  to go to top of page
                               ; BACK to go back one page
```

PROGRAM 17: DATA ENTRY

```
---DATA ENTRY FOR GRADUALLY VARIED FLOW WATER SURFACE DETERMINATION
   FOR PIPE---PAGE 2

Enter maximum number of intervals to be
used in profile.................................... ===>    "NN"
:ALLOWABLE VALUES ARE [10    ] TO [1000  ]

Enter constant pipe diameter(INCHES)............... ===>    "DIAM"
:ALLOWABLE VALUES ARE [3] TO [240   ]









--------------------------------------------------------------------------------
TYPE: EXIT to leave program ; TOP  to go to top of page
                            ; BACK to go back one page
```

PROGRAM 16

```
C  ------------------------------------------------------------------
      SUBROUTINE GVF(KTYPE)
C  ------------------------------------------------------------------
C
C
C  ANALYSIS OF GRADUALLY VARIED FLOW WATER
C  SURFACE DETERMINATION
C
      NT=NUT
C
      COMMON /NUT/NUT
C
C  INITIALIZE REQUIRED INPUT VARIABLES
      Z=0.0
      B=.001
      Q=0.0
      S0=0.0
      XL=0.
      RN=0.
      YCONT=0.
      NN=0
      DIAM=0.
C
C
C..READ DATA INPUT
      READ FREE(5)S0,XL,Q,RN,YCONT,NN,DIAM,B,Z
C
C
      IF(KTYPE.NE.8)GO TO 151
      CALL GVFPI(NT,DIAM,S0,Q,XL,RN,YCONT,NN)
      GO TO 153
151   CALL GVFCH(NT,Z,B,S0,Q,XL,RN,YCONT,NN)
153   CONTINUE
C
C
      RETURN
      END
```

```
C  -------------------------------------------------------------------
        SUBROUTINE GVFCH(NT,Z,B,S0,Q,XL,RN,YKNT,NN)
C  -------------------------------------------------------------------
C
C  ANALYSIS OF GRADUALLY VARIED FLOW WATER
C  SURFACE DETERMINATION
C
        COMMON/NUTP/NUTP,ISV
C
C  FUNCTION DEFINITIONS
        AREA(YY)=YY*(B+Z*YY)
        PER(YY)=B+2.*YY*SQRT(1.+Z*Z)
        YCTF(YY)=1.-Q*Q*(B+2.*Z*YY)/(G*AREA(YY)**3.)
        YNMF(YY)=1.-Q*Q*CON/(AREA(YY)**3.3333/PER(YY)**1.3333)
        DL(YY)=YCTF(YY)/(YNMF(YY)*S0)
        FPM(YY)=GAM*(YY*YY*(B*.5+Z*YY/3.)+Q*Q/(G*AREA(YY)))
        ENERGY(YY)=YY+Q*Q/(2.*G*AREA(YY)**2.)
C
C
        IF(B.LE.0.)B=.0001
CC------CONSTANTS
        GAM=62.4
        G=32.2
        CON=(RN/1.486)**2./S0
        WRITE(NT,187)
6       FORMAT(76('-'))
        WRITE(NT,200)
200     FORMAT(5X,'GRADUALLY VARIED FLOW PROFILE INPUT INFORMATION:')
        WRITE(NT,6)
        WRITE(NT,202)S0,XL,Q,RN,YKNT,NN,B,Z
202     FORMAT(5X,'CONSTANT CHANNEL SLOPE(FEET/FEET) = ',F8.6,/,
     C  5X,'CHANNEL LENGTH(FEET) = ',F12.2,/,
     C  5X,'CONSTANT CHANNEL FLOW(CFS) = ',F12.2,/,
     C  5X,'CONSTANT CHANNEL FRICTION FACTOR(MANNING) = ',F8.6,/,
     C  5X,'ASSUMED CHANNEL CONTROL DEPTH(FEET) = ',F8.2,/,
     C  5X,'MAXIMUM NUMBER OF INTERVALS IN PROFILE = ',I6,/,
     C  5X,'CONSTANT CHANNEL BASEWIDTH(FEET) = ',F10.2,/,
     C  5X,'CONSTANT CHANNEL "Z" FACTOR = ',F10.4)
C-------------------------------------------------------------
C       PROFILE DETERMINATION
C
        NN=NN*2
        UPIT=5000.
        DOWN=0.
        YC=1000.
        DO 520 I=1,22
        IF(YCTF(YC))514,521,515
514     DOWN=YC
        GOTO 520
515     UPIT=YC
520     YC=(UPIT+DOWN)*.5
521     IF(YKNT.EQ.0.)YKNT=YC
        UPIT=5000.
        DOWN=0.
        YN=1000.
        DO 530 I=1,22
        IF(YNMF(YN))523,531,524
523     DOWN=YN
        GOTO 530
```

```
524       UPIT=YN
530       YN=(UPIT+DOWN)*.5
531       CONTINUE
          WRITE(NT,533)YN,YC
533       FORMAT(5X,'NORMAL DEPTH(FEET) = ',F9.2,/,
     C    5X,'CRITICAL DEPTH(FEET) = ',F9.2)
535       IF(YN.LE.YC)GOTO 550

          IF(YKNT.LT.YC)GOTO 545
C
          SIGN=-1.
          DY=(YKNT-YN)*.998/NN
          KODE=1
          GOTO 560
545       SIGN=1.
          DY=(YC-YKNT)/NN
          KODE=2
          GOTO 560
C
550       IF(YKNT.LE.YC)GOTO 555
C
          SIGN=-1.
          DY=(YKNT-YC)/NN
          KODE=1
          GOTO 560
C
555       SIGN=1.
          DY=(YN-YKNT)*.998/NN
          KODE=2
560       SL=0.
          IF(ABS(DY).LT..0005)GOTO 1500
          Y=YKNT
          E=ENERGY(Y)
          FM=FPM(Y)
C
C------OUTPUT
          WRITE(NT,180)
          IF(KODE.EQ.1)WRITE(NT,262)YKNT
          IF(KODE.EQ.2)WRITE(NT,263)YKNT
262       FORMAT(5X,'DOWNSTREAM CONTROL ASSUMED DEPTH(FT) = ',F8.2)
263       FORMAT(5X,'UPSTREAM CONTROL ASSUMED DEPTH(FT) = ',F8.2)
C
          WRITE(NT,180)
          WRITE(NT,264)
264     FORMAT(5X,'GRADUALLY VARIED FLOW PROFILE COMPUTED INFORMATION',
     .    ':')
          WRITE(NT,6)
          WRITE(NT,261)
261       FORMAT(2X,'DISTANCE FROM',6X,'FLOWDEPTH',3X,'VELOCITY',6X,
     .    'SPECIFIC',8X,
     C    'PRESSURE+',/,3X,'CONTROL(FT)',9X,'(FT)',6X,'(FT/SEC)',
     .    5X,'ENERGY(FT)',
     C    4X,'MOMENTUM(POUNDS)')
          VV=SQRT((E-Y)*64.36)
          WRITE(NT,564)SL,Y,VV,E,FM
C-------PROFILE CALCULATION
564       FORMAT(2F15.3,F11.3,F15.3,2X,F15.2)
          LINE=6
          DO 580 I=1,NN,2
          Y2=YKNT+SIGN*DY*(I+1)
```

```
         DX=DY*(DL(Y)+DL(Y2)+4.*DL(YKNT+SIGN*I*DY))/3.
         SL=SL+DX
         IF(SL.GT.XL)GOTO 582
         Y=Y2
         E=ENERGY(Y)
         FM=FPM(Y)
         IF(I.EQ.NN-1.AND.SL.LT.0.)SL=XL
C
1580     VV=SQRT((E-Y)*64.36)
580      WRITE(NT,564)SL,Y,VV,E,FM
         GOTO 1000
C
C
582      Y=Y2-SIGN*2.*DY*(SL-XL)/DX
         E=ENERGY(Y)
         FM=FPM(Y)
         VV=SQRT((E-Y)*64.36)
         WRITE(NT,564)XL,Y,VV,E,FM
         GOTO 1600
1500     CONTINUE
         WRITE(NT,1505)
1505   FORMAT(/1X,'*****WARNING, PROFILE DEPTH INCREMENT IS ',
     .   'TOO SMALL:',/,
     C6X,'REATTEMPT PROBLEM WITH A SLIGHTLY DIFFERENT CONTROL DEPTH',/,
     C   6X,'OR A FEWER NUMBER OF PROFILE INTERVALS.')
1600     CONTINUE
1000     CONTINUE
C
C   FORMATS
C
180      FORMAT(76('='))
187      FORMAT(76('*'))
C
         RETURN
         END
```

PROGRAM 17

```
C  -----------------------------------------------------------------
         SUBROUTINE GVFPI(NT,DIAM,S0,Q,XL,RN,YKNT,NN)
C  -----------------------------------------------------------------
C
C  ANALYSIS OF GRADUALLY VARIED FLOW WATER
C  SURFACE DETERMINATION
C
         DELTA(YY)=ACOS((R-YY)/R)
         AREA(YY)=R*R*(ANG-.5*SIN(2.*ANG))
         PER(YY)=2.*R*ANG
         TW(YY)=2.*R*SIN(ANG)
         YCTF(YY)=1.-Q*Q*TW(YY)/(G*AREA(YY)**3.)
         YNMF(YY)=1.-Q*Q*CON/(AREA(YY)**3.3333/PER(YY)**1.3333)
         DL(YY)=YCTF(YY)/(YNMF(YY)*S0)
         FPM(YY)=GAM*(Q*Q/(G*AREA(YY))+YB*AREA(YY))
         ENERGY(YY)=YY+Q*Q/(2.*G*AREA(YY)**2.)
C
C
C  CONSTANTS
         GAM=62.4
         G=32.2
         CON=(RN/1.486)**2./S0
         WRITE(NT,187)
6        FORMAT(76('-'))
         WRITE(NT,200)
200      FORMAT(5X,'GRADUALLY VARIED FLOW PROFILE INPUT INFORMATION:')
         WRITE(NT,6)
         WRITE(NT,202)S0,XL,Q,RN,YKNT,NN,DIAM
202      FORMAT(5X,'CHANNEL SLOPE(FEET/FEET) = ',F8.6,/,
     .   5X,'CHANNEL LENGTH(FEET) = ',F12.2,/,
     C   5X,'CONSTANT CHANNEL FLOW(CFS) = ',F12.2,/,
     C   5X,'CONSTANT CHANNEL FRICTION FACTOR(MANNING) = ',F8.6,/,
     C   5X,'ASSUMED CHANNEL CONTROL DEPTH(FEET) = ',F8.2,/,
     C   5X,'MAXIMUM NUMBER OF INTERVALS IN PROFILE = ',I6,/,
     C   5X,'CONSTANT PIPE DIAMETER(INCHES) = ',F8.3)
C-----------------------------------------------------
C  PROFILE DETERMINATION
C
         LOGIC=0
         NN=2*NN
         DIAM=DIAM/12.
         R=DIAM/2.
         DMAX=.94*DIAM
512      UPIT=DIAM
         DOWN=0.
         YC=DMAX/2.
         DO 520 I=1,22
         ANG=DELTA(YC)
         IF(YCTF(YC))514,521,515
514      DOWN=YC
         GOTO 520
515      UPIT=YC
520      YC=(UPIT+DOWN)*.5
521      IF(YC.GT.DIAM)YC=DIAM
         IF(YKNT.EQ.0.)YKNT=YC
         TEST=.498*DIAM**2.6667*S0**.5/RN
         IF(Q.LT.TEST)GOTO 522
         WRITE(NT,5221)
5221     FORMAT(5X,'==>NORMAL PIPEFLOW IS PRESSURE FLOW')
```

```
            YN=DIAM*2.
            GOTO 531
      522   UPIT=DMAX
            DOWN=0.
            YN=DMAX/2.
            DO 530 I=1,22
            ANG=DELTA(YN)
            IF(YNMF(YN))523,531,524
      523   DOWN=YN
            GOTO 530
      524   UPIT=YN
      530   YN=(UPIT+DOWN)*.5
      531   CONTINUE
            DMAX=.82*DIAM
            IF(YN.LE.DIAM)WRITE(NT,533)YN,YC
      533   FORMAT(5X,'NORMAL DEPTH(FEET) = ',F9.2,/,
          C 5X,'CRITICAL DEPTH(FEET) = ',F9.2)
            IF(YN.LT.DIAM.AND.YN.GT.DMAX)WRITE(NT,5331)
      5331  FORMAT(5X,'NOTE:GIVEN NORMAL DEPTH IS LOWER VALUE OF TWO POSSIBLE.'
          C ,/,5X,'SUGGEST CONSIDERATION OF WAVE ACTION, UNCERTAINTY, ETC.')
            IF(YN.GT.DIAM)WRITE(NT,534)YC
      534   FORMAT(5X,'CRITICAL DEPTH(FEET) = ',F9.2)
      535   IF(YN.LE.YC)GOTO 550
      C
            IF(YKNT.LT.YC)GOTO 545
      C
            IF(YKNT.LT.YN)GOTO 5351
            IF(YN.GE.DMAX)LOGIC=1
            IF(LOGIC.EQ.1)GOTO 2000
            IF(YKNT.GT.DIAM)YKNT=DIAM
            GOTO 5359
      5351  IF(YN.GE.DMAX)LOGIC=2
            IF(LOGIC.EQ.2)GOTO 2000
      5359  SIGN=-1.
            DY=(YKNT-YN)*.998/NN
            KODE=1
            GOTO 560
      545   SIGN=1.
      C
            DY=(YC-YKNT)/NN
            KODE=2
      C
            GOTO 560
      C
      550   IF(YKNT.LE.YC)GOTO 555
      C
            IF(YN.GE.DMAX)LOGIC=4
            IF(LOGIC.EQ.4)GOTO 2000
            IF(YKNT.GT.DIAM)YKNT=DIAM
            SIGN=-1.
            DY=(YKNT-YC)/NN
            KODE=1
            GOTO 560
      C
      555   SIGN=1.
            IF(YKNT.LT.YN)GOTO 5559
            IF(YN.GE.DMAX)LOGIC=5
            IF(LOGIC.EQ.5)GOTO 2000
      5559  DY=(YN-YKNT)*.998/NN
            KODE=2
```

```
560     SL=0.
        IF(ABS(DY).LT..0005)GOTO 1500
        Y=YKNT
        ANG=DELTA(Y)
        E=ENERGY(Y)
        CALL YBAR(Y,DIAM,YB,NT)
        FM=FPM(Y)
C  OUTPUT
        WRITE(NT,180)
        IF(KODE.EQ.1)WRITE(NT,262)YKNT
        IF(KODE.EQ.2)WRITE(NT,263)YKNT
262     FORMAT(5X,'DOWNSTREAM CONTROL ASSUMED DEPTH(FT) = ',F8.2)
263     FORMAT(5X,'UPSTREAM CONTROL ASSUMED DEPTH(FT) = ',F8.2)
C
C
        WRITE(NT,180)
        WRITE(NT,264)
264     FORMAT(5X,'GRADUALLY VARIED FLOW PROFILE COMPUTED INFORMATION:')
        WRITE(NT,6)
        WRITE(NT,261)
261     FORMAT(2X,'DISTANCE FROM',6X,'FLOWDEPTH',3X,'VELOCITY',6X,
     .  'SPECIFIC',8X,
     C  'PRESSURE+',/,3X,'CONTROL(FT)',9X,'(FT)',6X,'(FT/SEC)',
     .  5X,'ENERGY(FT)',
     C  4X,'MOMENTUM(POUNDS)')
        VV=SQRT((E-Y)*64.36)
        WRITE(NT,564)SL,Y,VV,E,FM
C-------PROFILE CALCULATION
564     FORMAT(2F15.3,F11.3,F15.3,2X,F15.2)
C
C
        DO 580 I=1,NN,2
        Y2=YKNT+SIGN*DY*(I+1)
        ANG=DELTA(Y)
        TEMP1=DL(Y)
        ANG=DELTA(Y2)
        TEMP2=DL(Y2)
        ANG=DELTA(YKNT+SIGN*I*DY)
        TEMP3=DL(YKNT +SIGN*I*DY)
        DX=DY*(TEMP1+TEMP2+4.*TEMP3)/3.
        SL=SL+DX
        IF(SL.GT.XL)GOTO 582
        Y=Y2
        ANG=DELTA(Y)
        E=ENERGY(Y)
        CALL YBAR(Y,DIAM,YB,NT)
        FM=FPM(Y)
        IF(I.EQ.NN-1.AND.SL.LT.0.)SL=XL
C
C
1580    VV=SQRT((E-Y)*64.36)
580     WRITE(NT,564)SL,Y,VV,E,FM
C
C
        GOTO 1000
582     Y=Y2-SIGN*2.*DY*(SL-XL)/DX
        ANG=DELTA(Y)
        E=ENERGY(Y)
        CALL YBAR(Y,DIAM,YB,NT)
        FM=FPM(Y)
```

```
          VV=SQRT((E-Y)*64.36)
          WRITE(NT,564)XL,Y,VV,E,FM
1000      CONTINUE
          GOTO 2100
1500      CONTINUE
          WRITE(NT,1505)
1505      FORMAT(/1X,'*****WARNING,PROFILE DEPTH INCREMENT IS TOO SMALL:',
     C/6X,'REATTEMPT PROBLEM WITH A SLIGHTLY DIFFERENT CONTROL DEPTH',/,
     C 6X,'OR A FEWER NUMBER OF PROFILE INTERVALS.')
          GOTO 2100
2000      CONTINUE
          WRITE(NT,6)
          WRITE(NT,2005)
2005      FORMAT(5X,'FLOW PROFILE DEPTHS ARE GREATER THAN .82*DIAMETER.',
     C  /,5X,'SUGGEST CONSIDERATION OF SEALED FLOW IN THIS REACH DUE TO',/,
     C  5X,'WAVE ACTION, UNCERTAINTY, ETC.')
2100      CONTINUE
C
C  FORMATS
C
180       FORMAT(76('='))
187       FORMAT(76('*'))
C
          RETURN
          END
```

```
*********DESCRIPTION OF RESULTS*********************************************
*                                                                          *
*                    GRADUALLY VARIED FLOW IN A PIPE                       *
*                                                                          *
****************************************************************************

============================================================================
    GRADUALLY VARIED FLOW PROFILE INPUT INFORMATION:
----------------------------------------------------------------------------
    CHANNEL LENGTH(FEET) =       500.00
    CONSTANT CHANNEL SLOPE(FEET/FEET) =  .005000
    CONSTANT CHANNEL FRICTION FACTOR(MANNING) =  .013000
    MAXIMUM NUMBER OF INTERVALS IN PROFILE =     20
    CONSTANT CHANNEL FLOW(CFS) =       100.00
    ASSUMED CHANNEL CONTROL DEPTH(FEET) =      .00
    CONSTANT PIPE DIAMETER(INCHES) =   48.000
============================================================================
    NORMAL DEPTH(FEET) =      3.22
    CRITICAL DEPTH(FEET) =      3.03
============================================================================
    DOWNSTREAM CONTROL ASSUMED DEPTH(FT) =     3.03
----------------------------------------------------------------------------
    GRADUALLY VARIED FLOW PROFILE COMPUTED INFORMATION:
============================================================================
  DISTANCE FROM      FLOWDEPTH       SPECIFIC        PRESSURE+
   CONTROL(FT)          (FT)        ENERGY(FT)    MOMENTUM(POUNDS)
----------------------------------------------------------------------------
          .000         3.030          4.519          2765.09
          .095         3.039          4.519          2765.13
          .393         3.049          4.519          2765.24
          .919         3.059          4.519          2765.42
         1.702         3.068          4.520          2765.67
         2.777         3.078          4.520          2766.00
         4.184         3.088          4.521          2766.39
         5.975         3.098          4.522          2766.85
         8.213         3.107          4.523          2767.38

        10.977         3.117          4.524          2767.97
        14.370         3.127          4.525          2768.64
        18.526         3.136          4.526          2769.37
        23.629         3.146          4.527          2770.16
        29.933         3.156          4.529          2771.02
        37.812         3.165          4.530          2772.11
        47.844         3.175          4.532          2773.39
        60.994         3.185          4.534          2774.74
        79.061         3.194          4.536          2776.16
       106.052         3.204          4.538          2777.63
       154.556         3.214          4.540          2779.17
       500.000         3.222          4.542          2780.44
```

CHAPTER 4.

HYDRAULIC ELEMENTS II

Introduction

In this chapter, FORTRAN computer programs are presented for the analysis of backwater flow profiles in irregular channels, the hydraulic analysis of normal depth flow in irregular channels, and the modeling of flow-through detention basin effects. The programs are compatible with most of the currently available microcomputers which support the FORTRAN computer language. The engineer can extend the programs to accomodate large scale problems by simply modifying the various program memory array dimension statements.

PROGRAM 18. Irregular Channel Backwater Curve Analysis

Chapter 3 presents computer programs which include the capability for the analysis of backwater and drawdown water surface profiles in regular channels such as rectangular or trapezoidal sections, and pipelines. PROGRAM 18 extends the analysis procedure to irregular channels by use of the Standard Step Method (Chow, 1959).

The Standard Step Method provides for the computation of an upstream water surface and EGL given the current values of the water surface elevation, EGL, flow rate, and other flow and channel characteristics. Consequently, the main thrust of the computer program is to develop backwater and drawdown curves for channel flow situations which can be considered mild (i.e., Froude numbers are all less than 1, and the normal depths are greater than the critical depths).

The program data entry requirements fall into two categories:

(1) preparation of channel cross section information
(2) definition of uniform channel flowrate, energy balance locations along the channel, and downstream hydraulic control information

The preparation of channel cross section information entails the definition of up to 20 cross sections along the channel. Each cross section is defined to have up to 20 (x,y) coordinate pairs, a Manning's friction factor, a kinetic energy correction factor, and an eddy loss factor. Each cross section is assumed to have a single flowline such that the section begins on one bank, decreases in elevation constantly to the flowline, and then increases in elevation until the other bank elevation is reached. The section coordinate data is entered with the first x-coordinate being defined as x= 0.0. Thus the coordinate data must be entered consistently in order to scan the cross sections properly. That is, all sections should have the coordinate data prepared from left-to-right (or right-to-left). The cross section data entry begins with the most downstream section, with subsequent section data entered in the upstream direction.

The second set of data entry requirements includes the constant flowrate to be used through the entire study, the definition of the downstream hydraulic control water surface elevation, and the definition of energy balance locations along the channel. PROGRAM 18 computes the critical depth at each energy balance location. Consequently, should the specified control water surface elevation correspond to a channel flowdepth less than critical depth, the program will redefine the control to equal the critical depth. The Standard Step Method computes the water surface profile by balancing the energy losses (between energy balance locations) to the change in the EGL. Thus the engineer locates those points where the energy balance computations are to occur within the study.

As the study proceeds in the upstream direction, the upstream water surface and EGL are computed at energy balance locations by linearly interpolating all cross section geometric and hydraulic information. Should the water surface fall below the corresponding critical depth, the channel flowdepth is redefined to equal the critical depth (Froude number equals 1, modified by the kinetic energy correction factor). In reaches where normal depth flows could be supercritical, the word STEEP is noted in the right hand column of the computer solution tabulation. Other program output features are listed in the program description page which is included in the program output.

PROGRAM 18 is designed to tabulate the hydraulic computations according to the form given in Chow (1959, pg. 269). The computer results also include a report description page which briefly describes the computer program operation and the interpretation of the final results. It should be noted that PROGRAM 18 models the gradually varied flow hydraulics as a one-dimensional channel reach, ignoring all additional losses due to bends, angle points, and so forth. However, the eddy loss factor represents a portion of the velocity head to be used for energy losses, and can be used to include any other additional losses as a function of the flow velocity head, Hv. Should channel flows exceed either channel bank, the model assumes that a verticle line extends upwards from both channel banks, with zero friction assumed along the imaginary boundary. In this flooding situation, the word FLOOD is printed in the computer results.

PROGRAM 19. Irregular Channel Hydraulics

A common problem in flood control hydraulics is the estimation of normal depth flow hydraulics for an irregular channel which is composed of several subchannels. In PROGRAM 19, up to 20 channels can be defined, each with a different Manning's friction factor and cross section geometry. The subchannels are assumed to be connected as a global cross section through which the specified flowrate flows in equilibrium between the subchannels. That is, the flows are modeled as occuring in each subchannel such that the water surface elevation corresponds to each subchannel simultaneously.

The data entry requirements include the definition of each subchannel in terms of (x,y) coordinates. Each subchannel is assumed to have only one flowline and one friction factor. After the cross section data is entered, the global normal depth is computed by incrementally increasing the water surface elevation simultaneously in each subchannel until the corresponding normal depth flowrate exceeds the required total flowrate.

PROGRAM 20. Flow-Through Detention Basin Model (Modified Puls Method)

A motivation for inclusion of a stormwater detention basin in a flood control system is to reduce the runoff peak flow rates generated from the tributary watershed. The usual trade-off is to dedicate potentially developable land for detention basin purposes in exchange for a reduced flood control channel system downstream of the basin. Some of the advantages and disadvantages of use of detention basins are listed in Table 4.1.

The consideration of a detention basin system needs to address the various hydrologic, hydraulic, environmental and flood control concerns listed in Table 4.1. Of special concern is the interplay between several detention basins utilized in a flood control system. Improperly placed basins can increase the total downstream runoff peak above what would be normally generated without the basins. Due to detention basins decreasing the peak flow rate, the basin outflow peak flow rate is distributed over a longer duration and can sum directly with downstream basin outflow peak flow rates. This direct summation of basin outflow peak flow rates can have a serious impact in a watershed drainage system where the downstream areas are controlled by detention basins rather than being drained prior to the arrival of the upstream peak flow rate.

TABLE 4.1. DETENTION BASIN CONSIDERATIONS

Benefits	Potential Concerns
* reduces the peak rates of runoff to downstream reaches of the watershed	* detention basins do not reduce the total runoff volume unless basins are designed with dead storage (which is permanently stored within the basin)
* basins reduce the transport of sediments and debris carried by the floodwaters	* requires maintenance of storage capacity, inflow and outflow facilities
* inclusion of a detention basin in a flood control system may significantly reduce the downstream storm channel costs	* basins may discharge flows for long durations causing an increase in erosion
* provides a location for groundwater recharge	* improperly sized and placed basins may aggravate rather than reduce downstream flooding potential
* provides a location to concentrate floodwaters for contaminant treatment	* sediment and debris occupies flood control storage volume, reducing the basin benefit
	* maintenance costs for debris removal
	* detention basins may be unsightly in an urban area

Detention Basin Routing Procedure

The modified Puls method is frequently used for flow-through detention basin modeling purposes. The reservoir routing equation is based upon the conservation formula

$$I - O = dS/dT \tag{4.1}$$

where

I = detention basin inflow rate (cfs)
O = detention basin outflow rate (cfs)
dS = change in basin storage (cubic feet)
dT = time step (seconds)

Equation (4.1) can be approximated by replacing the variables I and O with time averaged values of

$$I = (I_1 + I_2)/2 \qquad (4.2)$$

$$O = (O_1 + O_2)/2 \qquad (4.3)$$

where the subscript 1 indicates the beginning of a time step, and the subscript 2 indicates the end of the timestep. Substituting the time averaged approximations for basin inflow and outflow into the conservation formula of (4.1) gives

$$(S_2 + O_2 dT/2) = (S_1 - O_1 dT/2) + (I_1 + I_2)dT/2 \qquad (4.4)$$

The right hand side of (4.4) is known from previously computed values of storage S_1, outflow O_1, and average basin inflow of the quantity $(I_1 + I_2)/2$ during the timestep dT.

Solution of the routing equation requires the following information:

(1). Initial conditions of basin storage and outflow
(2). A routing timestep, dT
(3). An inflow hydrograph (or unit runoff increments of duration dT)
(4). A functional relationship between basin storage, outflow, and basin depth

To aid in the calculation procedure, a storage indication curve may be developed from the storage-discharge relationships by selecting various values of outflow and computing (S + OdT/2). The solution procedure then consists of the following steps:

(1). Determine the average inflow volume from the inflow hydrograph and the selected unit period of dT by calculating the quantity $(I_1 + I_2)dT/2$.

(2). Determine $(S_1 - O_1 dT/2)$ from the initial condition of the basin, or from the last computed values of S and O.

(3). Use Puls equation (4.4) to compute $(S_2 + O_2 dT/2)$.

(4). Using the quantity determined in step 3, compute O_2 from the storage indication curve. The value O_2 is used to estimate the outflow volume during the timestep dT.

(5). Use O_2 to estimate S_2 from the basin storage versus outflow relationship.

These five steps are repeated for successive time steps using I_2, O_2, and S_2 as the new values of I_1, O_1, and S_1, respectively. The process is repeated until all the inflow values are used and the basin outflow becomes negligible. The above procedure is illustrated in the following example problem:

Example 4.1. Example Problem: Modified Puls Method

The assumed detention basin outflow and storage versus depth relationships are shown in Figs. 4.1 and 4.2, respectively. Using a timestep of dT = 60 minutes, the appropriate storage indication curve is developed in Table 4.2, and the resulting curve is plotted in Fig. 4.3. The example problem inflow hydrograph is tabulated in Table 4.3.

TABLE 4.2. EXAMPLE PROBLEM STORAGE INDICATION CURVE DEVELOPMENT

Depth (ft)	O (cfs)	S (AF)	S − OdT/2 (AF)	S + OdT/2 (AF)
0	0.0	0.0	0.0	0.0
1	4.2	14.4	14.22	14.57
2	12.0	28.8	28.30	29.30
3	51.7	43.2	41.06	45.34
4	114.7	57.6	52.86	62.34
5	186.8	72.0	64.28	79.72
6	263.2	86.4	75.52	97.28

Assuming an initial condition of zero basin outflow and storage, an example runoff hydrograph is routed using the formulation of (4.4) in Table 4.3. The salient features of the routed hydrograph are shown in Fig. 4.4.

TABLE 4.3. EXAMPLE PROBLEM BASIN ROUTING TABULATION

Time (min)	Inflow (cfs)	Mean Inflow	$(I_1+I_2)dT/2$ (AF)	$(S_1-O_1dT/2)$ (AF)	$(S_2+O_2dT/2)$ (AF)	Outflow (cfs)
0	0					0.0
		30	2.48	0.00	2.48	
60	60					0.7
		90	7.44	2.42	9.86	
120	120					2.8
		200	16.53	9.62	26.16	
180	280					10.3
		265	21.90	25.31	47.21	
240	250					58.6
		235	19.42	42.37	61.79	
300	220					112.7
		170	14.05	52.48	66.53	
360	120					132.1
		110	9.09	55.61	64.70	
420	100					124.5
		80	6.61	54.41	61.02	
480	60					109.8
		30	2.48	51.94	54.42	
540	0					85.4
		0	0.00	47.36	47.36	
600	0					59.2

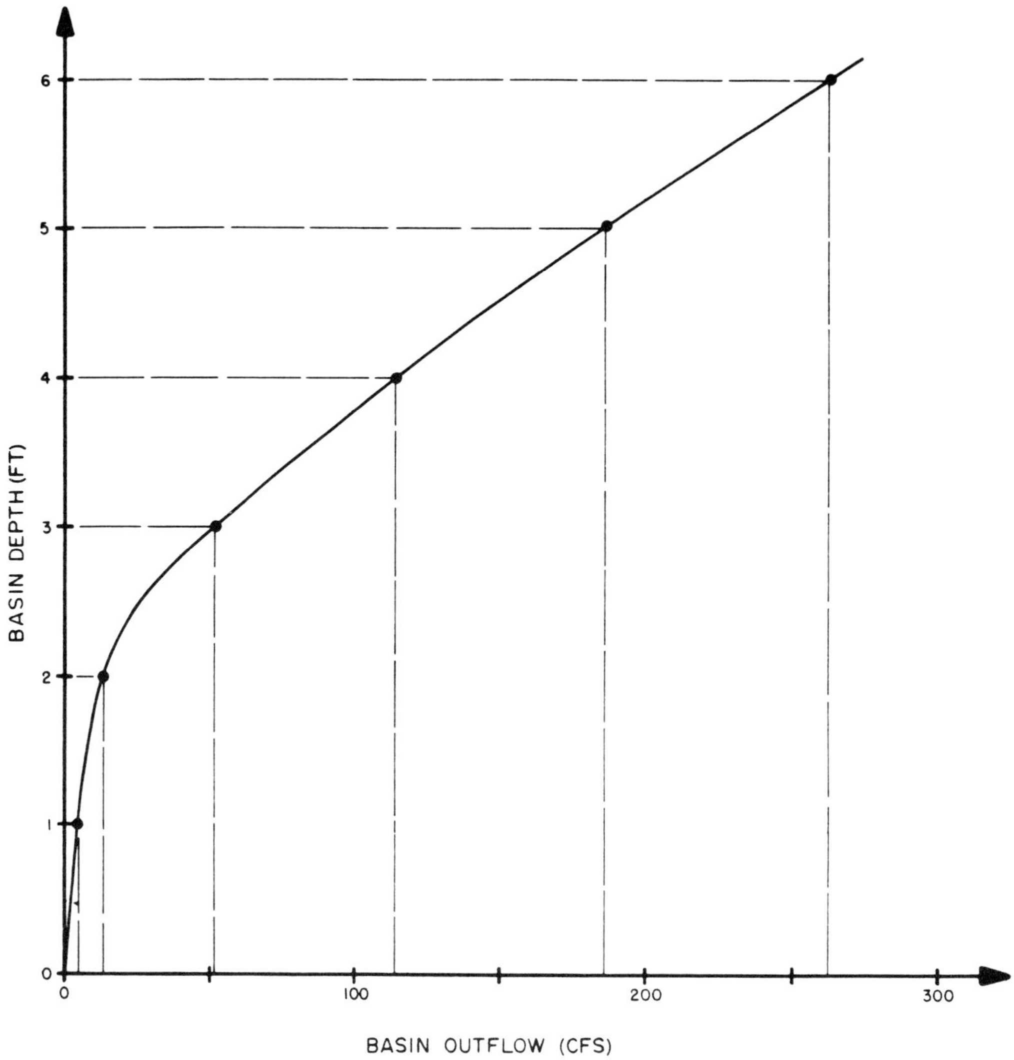

Fig. 4.1. Example problem detention basin outflow (cfs) as a function of depth (ft)

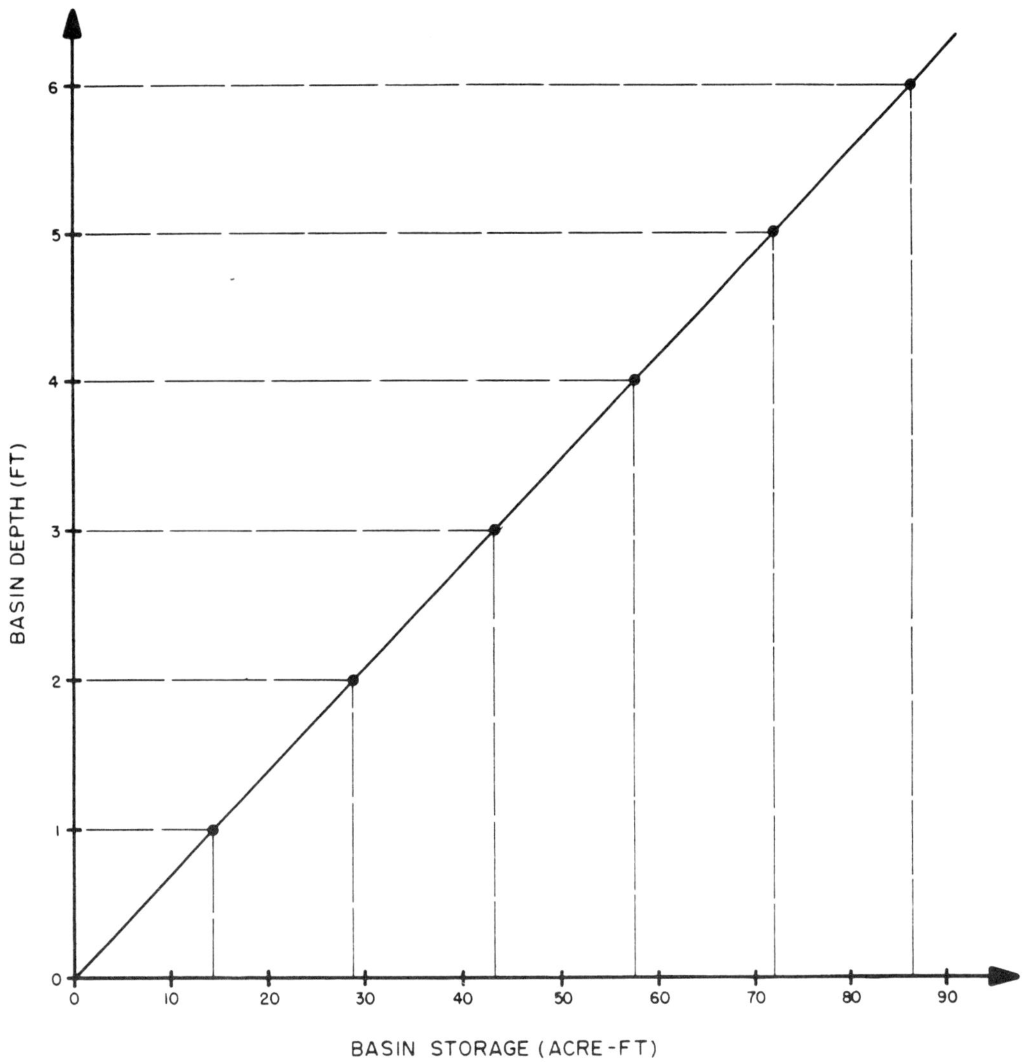

Fig. 4.2. Example problem detention basin storage (AF) as a function of depth (ft)

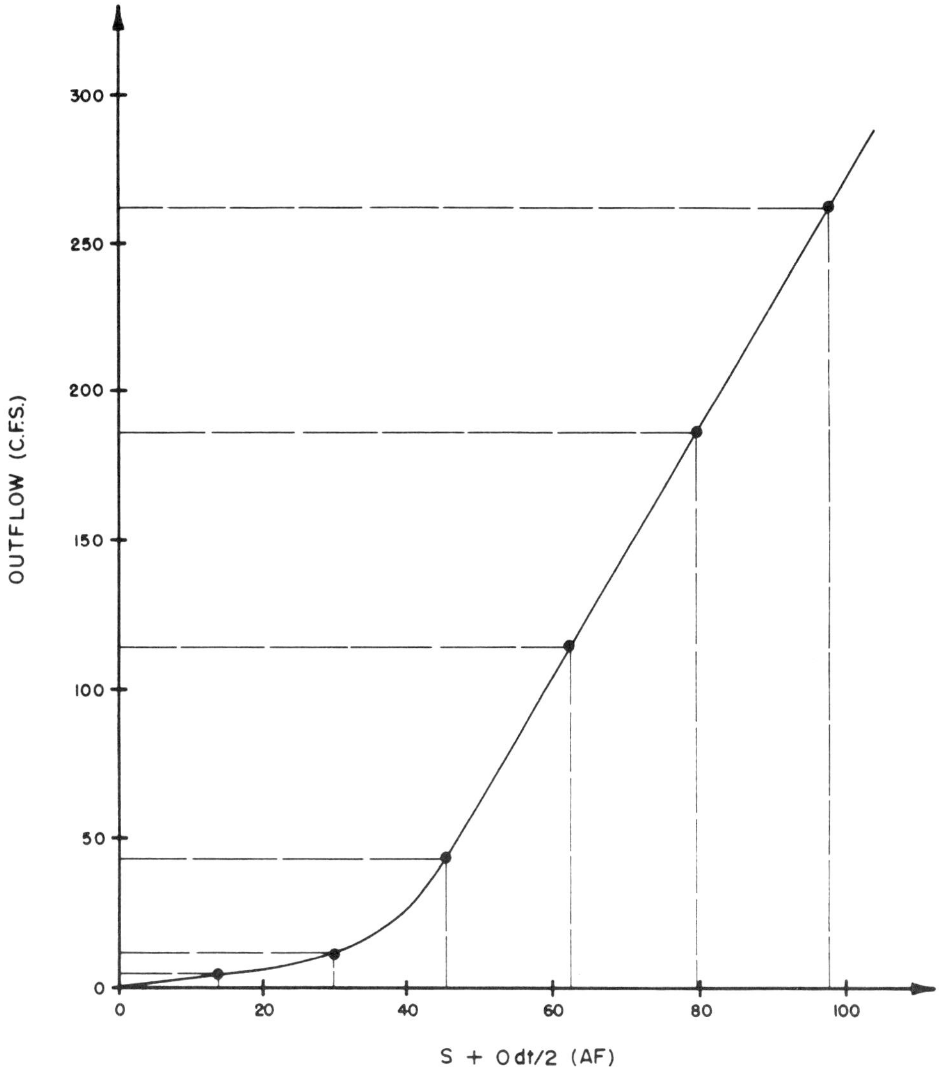

Fig. 4.3. *Example problem storage-indication curve*

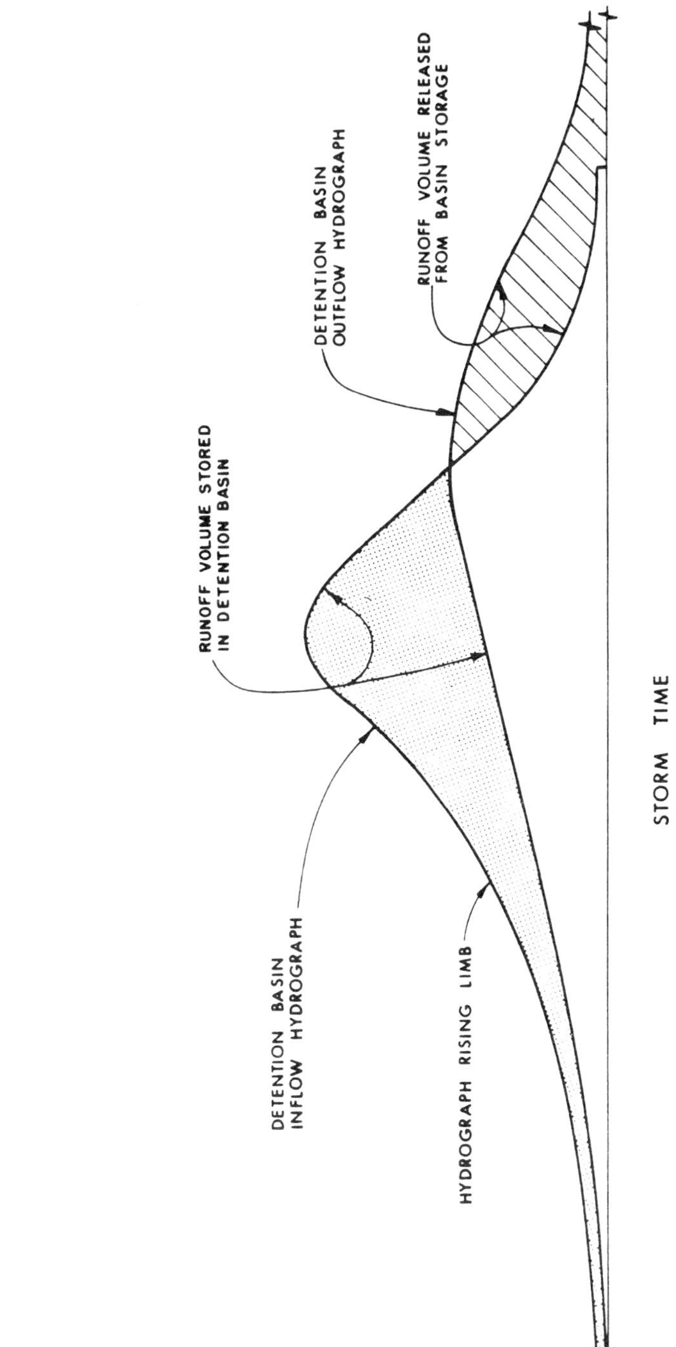

Fig. 4.4. Detention basin inflow and outflow hydrographs

Detention Basin Design Critical Storm Pattern

Generally, a 24-hour duration synthetic critical storm pattern will adequately test a detention basin for a selected return frequency of rainfall event. Such a synthetic runoff hydrograph method is developed in Chapter 5, along with a FORTRAN computer program package for its implementation on a microcomputer. A synthetic runoff hydrograph is developed and used as the detention basin inflow hydrograph, and the previous basin routing formulation used to test the basin response to the severe design storm conditions. In most cases, detention basin design will attenuate the inflow hydrograph peak flow rate by less than about 15 percent and the single 24-hour storm pattern may be adequate for the flow-through basin size determination. For larger percentages of peak flow reduction, or for detention basins where significant storage effects are attributed towards dead storage (permanent storage of floodwaters and debri with percolation being the only outlet for the floodwaters), a longer duration critical storm pattern may be required to adequately test the basin design capacity.

One approach for extending the critical storm pattern duration is to include additional 24-hour durations of storm rainfall with the total series of storms representing the point rainfall data with the selected return frequency. If record rain gage data is unavailable, peak 48 hours, 72 hours, etc. of critical storm mass rainfall can be estimated by logarithmic extrapolation of the known mass rainfall values. For example, assuming there exists a straight line relationship between duration and mass rainfall, the computed area-averaged 6- and 24-hour point rainfalls can be plotted on standard log-log paper and the longer duration mass rainfalls estimated from the resulting straight line plotting. Using the mass rainfall plot, second day (or longer) mass rainfall values can be estimated and the associated storm pattern determined from a direct proportion of the peak 24- hour storm pattern. The resulting second day storm pattern is then placed to precede the peak 24-hour storm pattern such as shown in Fig. 4.5 for a two day storm sequence. In this fashion, a severe test of the detention basin is achieved using local point rainfall information with the desired return frequency.

Chapter Bibliography

Chow, Ven Te, Open-Channel Hydraulics, McGraw-Hill, New York, 1959.

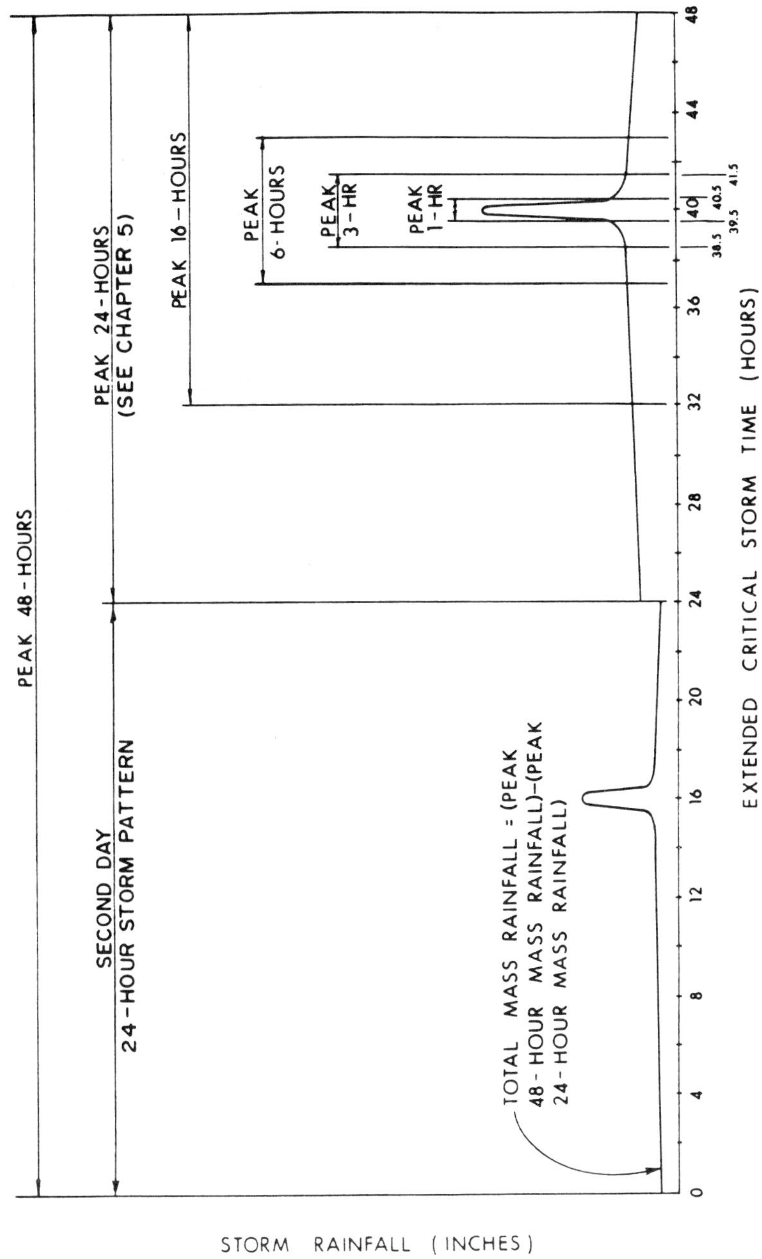

Fig. 4.5. Extended synthetic critical storm

PROGRAM 18: DATA ENTRY

```
---DATA ENTRY FOR IRREGULAR CHANNEL WATER SURFACE PROFILE---PAGE 1

In order to determine a water surface profile in a sub-
critical channel, the channel needs to be defined by
cross-sections. These sections are entered as one tra-
vels upstream.
Enter TOTAL NUMBER of cross-sections................ ===>    "N"
:ALLOWABLE VALUES ARE [2] TO [20    ]

Enter TOTAL irregular channel FLOW(CFS)............. ===>    "QQ"
:ALLOWABLE VALUES ARE [1     ] TO [999999]

Enter allowable ERROR(FEET) in water surface
determination....................................... ===>    "TOL"
:ALLOWABLE VALUES ARE [.01   ] TO [.50   ]

-------------------------------------------------------------------------------
    TYPE: EXIT to leave program ; TOP  to go to top of page
```

```
---DATA ENTRY FOR IRREGULAR CHANNEL WATER SURFACE PROFILE---PAGE 2


Enter hydraulic control water surface(ELEVATION)
at cross-section  1 ................................ ===>    "YCON"
:ALLOWABLE VALUES ARE [-9999 ] TO [9999  ]
(NOTE: WATER SURFACE ELEVATION IS WITH RESPECT
       TO CROSS-SECTION £1 ELEVATION/COORDINATES)

-------------------------------------------------------------------------------
    TYPE: EXIT to leave program ; TOP  to go to top of page
                                ; BACK to go back one page
```

```
---DATA ENTRY FOR IRREGULAR CHANNEL WATER SURFACE PROFILE---PAGE 3

CHANNEL CROSS-SECTION DATA ENTRY GROUP  1 OF  2:

Enter TOTAL number of nodal points to describe
cross-section...................................... ===>    "M(I)"
:ALLOWABLE VALUES ARE [3] TO [20    ]
(NOTE: INCLUDE BOTH END POINTS IN TOTAL NUMBER)

Enter Mannings friction factor..................... ===>    "XMAN(I)"
:ALLOWABLE VALUES ARE [.008  ] TO [.99   ]

Enter kinetic energy correction factor............. ===>    "ALPHA(I)"
:ALLOWABLE VALUES ARE [1.0   ] TO [2.0   ]
(NOTE: THIS CROSS-SECTION FACTOR INCREASES THE
VELOCITY HEAD TERM IN ORDER TO ACCOUNT FOR VARIATIONS
IN VELOCITY FROM THE AVERAGE FLOW VELOCITY.  GENERALLY,
[1.1] IS AN ADEQUATE VALUE FOR TYPICAL CHANNEL FLOWS.)


--------------------------------------------------------------------------------
    TYPE: EXIT to leave program ; TOP  to go to top of page
                                ; BACK to go back one page
```

```
---DATA ENTRY FOR IRREGULAR CHANNEL WATER SURFACE PROFILE---PAGE 3A

CHANNEL CROSS-SECTION DATA ENTRY GROUP  1 OF  2:

Enter eddy loss coefficient......................... ===>    "EDDY(I)"
:ALLOWABLE VALUES ARE [0.0   ] TO [3.0   ]
(NOTE: THIS VALUE IS USED TO APPROXIMATE ENERGY LOSSES
 DUE TO CHANGES IN THE CHANNEL SHAPE AND OTHER RELATED
 LOSSES.  GENERALLY, THE MORE IRREGULAR THE CHANNEL,
 THE HIGHER THE EDDY LOSS TERM.  THE PROGRAM USER MAY
 ALSO INCLUDE OTHER ENERGY LOSSES IN THE ANALYSIS BY
 SELECTING ON EQUIVALENT EDDY LOSS.)

 FOR SUBCRITICAL FLOW ANALYSIS, A HYDRAULIC CONTROL MUST EXIST
 AT THE MOST DOWNSTREAM POINT OF THE CHANNEL BEING STUDIED.
 EACH CHANNEL CROSS-SECTION IS LOCATED WITHIN THE CHANNEL BY ITS
 DISTANCE(FEET) FROM THE CHANNEL CONTROL.  SINCE CROSS-SECTION
 ONE IS THE MOST DOWNSTEAM SECTION, THE PROGRAM ASSUMES THAT THE
 CONTROL OCCURS AT THE LOCATION.  ALL CHANNEL LOCATIONS WILL BE
 REFERENCED WITH RESPECT TO SECTION NUMBER ONE.

--------------------------------------------------------------------------------
TYPE: EXIT to leave program ; TOP  to go to top of page
                            ; BACK to go back one page
```

```
---DATA ENTRY FOR IRREGULAR CHANNEL WATER SURFACE PROFILE---PAGE 3B
             CROSS-SECTION   1 NODAL POINT  2 OF  3:

Enter (X) coordinate................................ ===>    "X(I,J)"
:ALLOWABLE VALUES ARE [        .01] TO [9999       ]

Enter (Y) elevation................................. ===>    "Y(I,J)"
:ALLOWABLE VALUES ARE [-9999      ] TO [9999       ]

NODE  (X) COORDINATE     (Y) ELEVATION
  1       [    .00]          [  20.00]
          [       ]          [       ]
          [       ]          [       ]
          [       ]          [       ]
          [       ]          [       ]
          [       ]          [       ]
          [       ]          [       ]
          [       ]          [       ]
          [       ]          [       ]
          [       ]          [       ]

--------------------------------------------------------------------------------
TYPE: EXIT to leave program ; TOP  to go to top of page
                            ; BACK to go back one entry
```

```
---DATA ENTRY FOR IRREGULAR CHANNEL WATER SURFACE PROFILE---PAGE 3A

CHANNEL CROSS-SECTION DATA ENTRY GROUP  2 OF  2:

Enter eddy loss coefficient......................... ===>    "EDDY(I)"
:ALLOWABLE VALUES ARE [0.0   ] TO [3.0   ]




Enter distance(FEET) from assumed hydraulic control
section............................................ ===>    "DISTS(I)"
:ALLOWABLE VALUES ARE [       1.00] TO [  10000.00]



--------------------------------------------------------------------------------
TYPE: EXIT to leave program ; TOP  to go to top of page
                            ; BACK to go back one page
```

```
---DATA ENTRY FOR IRREGULAR CHANNEL WATER SURFACE PROFILE---PAGE 4

This computer program uses straight line interpolation between
cross-sections to define the total irregular channel.  Conse-
quently, channel shape,slope,friction, and other factors are
all averaged with respect to distance between cross-sections.
For water-surface profile determination, the locations where
the hydraulic energy balances are to be made is important to
the correct solution.

On the next page, the user is requested to enter the locations
where the hydraulic energy balances are to be made.  These
locations may be specified at or between the various channel
cross-sections previously entered.  Again  only subcritical
flow is analysed by this program, and critical-depth will be
assumed at all sections where subcritical flow occurs.

Enter total number of locations where hydraulic energy
balances are to be made(STANDARD STEP METHOD)....... ===>    "NE"
:ALLOWABLE VALUES ARE [5] TO [50    ]

--------------------------------------------------------------------------------
TYPE: EXIT to leave program ; TOP  to go to top of page
```

```
---DATA ENTRY FOR IRREGULAR CHANNEL WATER SURFACE PROFILE---PAGE 5

HYDRAULIC ENERGY BALANCE LOCATION  1 OF  5:

Enter distance(FEET) from hydraulics control to where
energy balance is to be made........................ ===>    "DIST(I)"
:ALLOWABLE VALUES ARE [      1.00] TO [      6.00]

                HYDRAULIC ENERGY
 LOCATION     BALANCE DISTANCES
                  [         ]
                  [         ]
                  [         ]
                  [         ]
                  [         ]
                  [         ]
                  [         ]
                  [         ]
                  [         ]
                  [         ]

--------------------------------------------------------------------------------
TYPE: EXIT to leave program ; TOP  to go to top of page
                            ; BACK to go back one page
```

PROGRAM 18

```
C  ------------------------------------------------------------
         SUBROUTINE PROF1E
C  ------------------------------------------------------------
C
C  DATA ENTRY ROUTINE FOR PROFILE BATCH SYSTEM
C
         COMMON/NUT/NUT
         COMMON/BLK1/X(20,20),Y(20,20),NFL(20),M(20)
         COMMON/BLK2/DISTS(20),ALPHA(20),XMAN(20),EDDY(20)
         COMMON/BLK3/TOL,QQ,N,DIST(50),NE
C
C
         NUT=6
C
         OPEN 5,"PROF.DAT"
         OPEN 6,"PROF.ANS"
C
C
C..READ DATA INPUT
         READ FREE(5) N,QQ,TOL,YCON
         DO 10 I=1,N
         READ FREE(5) M(I),XMAN(I),ALPHA(I),EDDY(I),DISTS(I)
         MM=M(I)
         DO 20 J=1,MM
         READ FREE(5) X(I,J),Y(I,J)
20       CONTINUE
10       CONTINUE
C
         READ FREE(5) NE
         DO 30 I=1,NE
         READ FREE(5) DIST(I)
30       CONTINUE
C
C..END OF INPUT
C
C
         CALL PROF5(NUT)
         CALL PROF6(NUT)
         CALL PROF2(NUT,YCON)
C
         RETURN
         END
C
C
C  ------------------------------------------------------------
         SUBROUTINE PROF2(NT,YCON)
C  ------------------------------------------------------------
C
C
         INTEGER BLANK
         COMMON/TEXTF/NOTE,BLANK,IST(22),IWT(12),IWTS(5)
         COMMON/BLK 1/X(20,20),Y(20,20),NFL(20),M(20)
         COMMON/BLK 2/DISTS(20),ALPHA(20),XMAN(20),EDDY(20)
         COMMON/BLK 3/TOL,Q,N,DIST(50),NE
C
C
         IFF=12
C
         IFIRST=0
```

```
         ICOUNT=1
         DCON=YCON
         ITYPE=BLANK
         CALL PROF4(0,YC,YN,J,K,NT,A,V,R,FR,ERROR)
         IF(YC.GE.YCON)DCON=YC
         IF(YC.GE.YCON)ITYPE=NOTE
         IF(YC.GE.YCON)GOTO 10
         CALL PROF3(1,YCON,A,V,P,DH,TW,R,FR,1.,NT,ERROR)
10       WRITE(NT,1000)IFF
1000     FORMAT(1X,A1,//,50X,'CUT ALONG OUTSIDE BORDER* *',/,
     C   76X,'*',/,76X,'*',/,76X,'*',/,76X,'*',/,76X,'V')
         WRITE(NT,1001)
1001     FORMAT(10X,'+',109('-'),'+')
         WRITE(NT,1002)
1002     FORMAT(2(10X,'|',109X,'|',/),10X,'|',109X,'|',/,
     .   10X,'|*** IRREGULAR CHANNEL SUBCRITICAL FLOW MODEL ***',7X,
     .   '(C) Copyright 1983 Advanced Engineering Software [AES]',
     .   '|',/,
     C   10X,'|Standard Step Method irregular channel',
     C   ' analysis. Based on development in "OPEN CHANNEL'
     C   ' HYDRAULICS",CHOW(1959)|')
         WRITE(NT,1010)Q
1010     FORMAT(10X,'|  STUDY NAME:',20X,'Channel Flow = ',F9.2,' cfs',
     C   16X,'PAGE NUMBER:',20X,'|')
         WRITE(NT,1001)
         WRITE(NT,1003)
1003     FORMAT(10X,'| LENGTH| WATER | FLOW|  FLOW | FLOW|',
     C   '  2   |  TOTAL | HYDR |FRICTION| AVERAGE| REACH|',
     C   ' LOSS|EDDY|  TOTAL |      |',/,10X,'|  from |SURFACE|',
     C   'DEPTH|  AREA |  V   |aV /2g|  HEAD  |RADIUS|  SLOPE |',
     C   '  REACH |LENGTH|  Hf |LOSS|  HEAD  |  Fr |',/,
     C   10X,'|CONTROL|(elev.)| (ft)|(ft*ft)|(fps)| (ft)  |  (ft)  |',
     C   ' (ft) |   Sf   |   Sf   | (ft) | (ft)|(ft)|  (ft)  |     |')
         WRITE(NT,1004)
1004     FORMAT(10X,'|-------|-------|-----|-------|-----|',
     C   '------|--------|------|--------|--------|------|',
     C   '-----|----|--------|-----|')
C
         SF=XMAN(1)*XMAN(1)*V*V/2.2082/R**1.3333
         AHV=ALPHA(1)*V*V/64.4
         H1=DCON+AHV
         H2=H1
         NODE=NFL(1)
         YY=DCON-Y(1,NODE)
         WRITE(NT,1005)DISTS(1),DCON,YY,A,V,AHV,H1,R,SF,H2,ITYPE,FR
1005     FORMAT(10X,'|'F7.1,'|',F7.2,'|',F5.2,'|',F7.1,'|',
     C   F5.2,'|',F6.3,'|',F8.3,'|',F6.2,'|',F8.6,'|        |',
     C   '     |    |    |',F8.3,'|',A1,F4.2,'|')
         DO 13 IJ=1,5
13       IWTS(IJ)=BLANK
         DO 131 IJ=1,12
131      IWT(IJ)=BLANK
         IF(ERROR.EQ.0.)GOTO 15
         DO 14 IJ=1,5
14       IWT(IJ)=IST(IJ)
         DO 141 IJ=6,12
141      IWT(IJ)=IST(10+IJ)
15       CONTINUE
         IF(YN.GT.YC)GOTO 17
         DO 16 IJ=1,5
```

```
16         IWTS(IJ)=IST(IJ+5)
           IF(IFIRST.NE.0)GOTO 79
           DO 78 IJ=1,5
78         IWTS(IJ)=IST(IJ+10)
79         IFIRST=IFIRST+1
17         WRITE(NT,1011)(IWT(IJ),IJ=1,5),(IWT(IJ),IJ=6,12),ALPHA(1),
      C    XMAN(1),(IWTS(IJ),IJ=1,5)
1011       FORMAT(10X,'|',1X,5A1,1X,'|',7A1,'|'5X,'|',7X,'|',5X,'|a=',
      C    F4.2,'|',8X,'|',6X,'|n=',F6.4,'|',8X,'|',6X,'|',
      C    5X,'|',4X,'|',8X,'|',5A1,'|')
           WRITE(NT,1004)
C.......INITIALIZE SF1
           SF1=SF
C.......MAIN LOOP
           DO 100 L=1,NE
           CALL PROF4(L,YC,YN,J,K,NT,A,V,R,FR,ERROR)
           YMIN=YC
           YMAX=YC+100.
           R=(DIST(L)-DISTS(J))/(DISTS(K)-DISTS(J))
           R1=1.-R
           JFL=NFL(J)
           KFL=NFL(K)
           YMINJ=Y(J,JFL)
           YMINK=Y(K,KFL)
           YFL=R*YMINK+R1*YMINJ
C.......ESTIMATE SECTION BANK-ELEVATIONS
           XLBNK=R*Y(K,1)+R1*Y(J,1)
           MMJ=M(J)
           MMK=M(K)
           XRBNK=R*Y(K,MMK)+R1*Y(J,MMJ)
           DO 50 LL=1,15
           YTEST=.5*(YMIN+YMAX)
           DY=YTEST-YFL
           YJ=YMINJ+DY
           YK=YMINK+DY
           CALL PROF3(J,YJ,AJ,V,PJ,DHJ,TWJ,XR,FR,1.,NT,ERROR)
           CALL PROF3(K,YK,AK,V,PK,DHK,TWK,XR,FR,1.,NT,ERROR)
           A=R*AK+R1*AJ
           V=Q/A
           AL=R*ALPHA(K)+R1*ALPHA(J)
           XN=R*XMAN(K)+R1*XMAN(J)
           AHV=AL*V*V/64.36
           TW=R*TWK+R1*TWJ
           P=R*PK+R1*PJ
           RH=A/P
           DH=A/TW
           SF=XN*XN*V*V/2.22/RH**1.3333
           DX=DIST(1)
           IF(L.GT.1)DX=DIST(L)-DIST(L-1)
           SFM=.5*(SF1+SF)
           HF=DX*SFM
           RDXE=DX/2./(DISTS(K)-DISTS(J))
           ED=(R-RDXE)*EDDY(K)+(R1+RDXE)*EDDY(J)
           HE=ED*V*V/64.36
           H2T=H2+HE+HF
           H1T=YTEST+AHV
           IF(H2T-H1T)30,60,40
30         YMAX=YTEST
           GOTO 50
40         YMIN=YTEST
```

```
50      CONTINUE
60      H1=H1T
        H2=H1
C.......OUTPUT RESULTS
        DO 61 IJ=1,5
61      IWTS(IJ)=BLANK
        DO 611 IJ=1,12
611     IWT(IJ)=BLANK
        IF(YTEST.LT.XLBNK.AND.YTEST.LT.XRBNK)GOTO 70
        DO 62 IJ=1,5
62      IWT(IJ)=IST(IJ)
70      CONTINUE
        FR=SQRT(AL*V*V/DH/32.18)
        ITYPE=BLANK
        TEST=ABS(1.-FR)
        IF(TEST.LT..01)ITYPE=NOTE
        SF1=SF
        IF(YN.GT.YC)GOTO 77
        DO 75 IJ=1,5
75      IWTS(IJ)=IST(IJ+5)
77      WRITE(NT,1006)DIST(L),YTEST,DY,A,V,AHV,H1,RH,SF,SFM,DX,HF,
     C  HE,H2T,ITYPE,FR
1006    FORMAT(10X,'|',F7.1,'|',F7.2,'|',F5.2,'|',F7.1,'|',
     C  F5.2,'|',F6.3,'|',F8.3,'|',F6.2,'|',F8.6,'|',F8.6,'|',
     C  F6.1,'|',F5.3,'|',F4.2,'|',F8.3,'|',A1,F4.2,'|')
C........SEE IF LOCATION =CROSS-SECTION
        IF(DIST(L).EQ.DISTS(J).OR.DIST(L).EQ.DISTS(K))GOTO 82
        GOTO 85
82      DO 83 IJ=6,12
83      IWT(IJ)=IST(10+IJ)
85      CONTINUE
        WRITE(NT,1012)(IWT(IJ),IJ=1,5),(IWT(IJ),IJ=6,12),
     C  AL,XN,ED,(IWTS(IJ),IJ=1,5)
1012    FORMAT(10X,'|',1X,5A1,1X,'|',7A1,'|',5X,'|',7X,'|',5X,'|a=',
     C  F4.2,'|',8X,'|',6X,'|n=',F6.4,'|',8X,'|',6X,'|',5X,
     C  'e=',F4.2,8X,'|',5A1,'|')
        WRITE(NT,1004)
C.......PAGE COSMETICS
        ICOUNT=ICOUNT+1
        IF(ICOUNT.LT.13)GOTO 100
        WRITE(NT,1001)
        IF(NE.EQ.L)GOTO 200
        WRITE(NT,1000)
        WRITE(NT,1001)
        WRITE(NT,1002)
        WRITE(NT,1010)Q
        WRITE(NT,1001)
        WRITE(NT,1003)
        WRITE(NT,1004)
        ICOUNT=0
100     CONTINUE
C.......END OF STUDY, BUT IN MIDDLE OF PAGE....FINISH OFF PAGE
        NREM=13-ICOUNT
        DO 150 K=1,NREM
        WRITE(NT,1007)
        WRITE(NT,1007)
1007    FORMAT(10X,'|',7X,'|',7X,'|',5X,'|',7X,'|',5X,'|',6X,
     C  '|',8X,'|',6X,'|',8X,'|',8X,'|',6X,'|',5X,'|',4X,'|',
     C  8X,'|',5X,'|')
        WRITE(NT,1004)
```

```
150     CONTINUE
        WRITE(NT,1001)
200     CONTINUE
        WRITE(NT,2000)
2000    FORMAT(/)
C
        RETURN
        END
C
C
C  --------------------------------------------------------------------
        SUBROUTINE PROF3(L,YDEPTH,A,V,P,DH,TW,R,FR,S,NT,ERROR)
C  --------------------------------------------------------------------
C
C
        COMMON/BLK 1/X(20,20),Y(20,20),NFL(20),M(20)
        COMMON/BLK 2/DISTS(20),ALPHA(20),XMAN(20),EDDY(20)
        COMMON/BLK 3/TOL,Q,N,DIST(50),NE
C
C
        HYPOT(B,C)=((B*B)+(C*C))**.5
        ERROR=0.
        I=L
        P=0.
        V=0.
        TW=0.
        A=0.
        KK=M(I)-1
        DO 200 J=1,KK
        IF((YDEPTH.LE.Y(I,J)).AND.(YDEPTH.LE.Y(I,(J+1))))GOTO 200
        XDIF=X(I,(J+1))-X(I,J)
        YDIF=Y(I,(J+1))-Y(I,J)
        IF(YDIF)150,150,160
150     YDIF=-YDIF
        YUP=Y(I,J)
        YLOW=Y(I,(J+1))
        GOTO 170
160     YUP=Y(I,(J+1))
        YLOW=Y(I,J)
170     TEMP=YUP-YDEPTH
        IF(TEMP.LE.0.)GOTO 155
        GOTO 165
155     AA=((YDEPTH-YUP)*XDIF)+(.5*XDIF*YDIF)
        PP=HYPOT(XDIF,YDIF)
        TWTW=XDIF
        GOTO 198
165     IF(ABS(YDIF).LT..001)GOTO 155
        YDEEP=YDEPTH-YLOW
        TWTW=XDIF*(YDEEP/YDIF)
        AA=.5*YDEEP*TWTW
        PP=(YDEEP/YDIF)*HYPOT(XDIF,YDIF)
198     A=A+AA
        P=P+PP
        TW=TW+TWTW
200     CONTINUE
        IF(P.LE..001)WRITE(NT,220)
220     FORMAT(1X,'UNACCEPTABLE CHANNEL CONFIGURATION')
        IF(P.LE..001)GOTO 180
        R=A/P
        V=Q/A
```

```
      DH=A/TW
      FR=SQRT(ALPHA(I)*V*V/32.18/DH)
180   CONTINUE
C.......CHECK FOR OVERFLOW
      KK=KK+1
      IF(YDEPTH.GT.Y(L,1).OR.YDEPTH.GT.Y(L,KK))ERROR=1.
C
      RETURN
      END
C
C
C ----------------------------------------------------------------------
      SUBROUTINE PROF4(I,YC,YN,J,K,NT,A,V,RH,FR,ERROR)
C ----------------------------------------------------------------------
C
C
C  DETERMINES RELATIVE DC ELEVATION AND SURROUNDING
C  SECTIONS "J" AND "K"
C  I IS INDEX TO ENERGY-BALANCE LOCATION
C  I=0 IMPLIES CONTROL SECTION ANALYSIS
C
C  FIND J AND K:
C
C
      COMMON/BLK 1/X(20,20),Y(20,20),NFL(20),M(20)
      COMMON/BLK 2/DISTS(20),ALPHA(20),XMAN(20),EDDY(20)
      COMMON/BLK 3/TOL,Q,N,DIST(50),NE
C
C
      DD=.01
      L=2
      IF(I.EQ.0)GOTO 15
      IF(I.GT.0)DD=DIST(I)
      DO 10 L=1,N
      IF(DD.GT.DISTS(L))GOTO 10
      GOTO 15
10    CONTINUE
15    J=L-1
      K=L
C.......GET LENGTH RATION R
      R=(DD-DISTS(J))/(DISTS(K)-DISTS(J))
      R1=1.-R
C.......FIND YC
      JFL=NFL(J)
      KFL=NFL(K)
      YMINJ=Y(J,JFL)
      YMINK=Y(K,KFL)
      YMIN=R*YMINK+R1*YMINJ
      YFL=YMIN
      YMAX=YMIN+100.
      AL=R*ALPHA(K)+R1*ALPHA(J)
      DO 100 L=1,15
      YTEST=.5*(YMIN+YMAX)
      DY=YTEST-YFL
      YJ=YMINJ+DY
      YK=YMINK+DY
      CALL PROF3(J,YJ,AJ,V,PJ,DHJ,TWJ,XR,FR,1.,NT,ERROR)
      CALL PROF3(K,YK,AK,V,PK,DHK,TWK,XR,FR,1.,NT,ERROR)
      A=R*AK+R1*AJ
      TW=R*TWK+R1*TWJ
```

```
      P=R*PK+R1*PJ
      RH=A/P
      DH=A/TW
      V=Q/A
      FR=(AL*V*V/32.18/DH)
      IF(FR-1.)20,120,40
20    YMAX=YTEST
      GOTO 100
40    YMIN=YTEST
100   CONTINUE
120   YC=YTEST
      FR=SQRT(FR)
C.......GET YN
      XN=R*XMAN(K)+R1*XMAN(J)
      SLOPE=(YMINK-YMINJ)/(DISTS(K)-DISTS(J))
      YN=0.
      IF(I.EQ.0)GOTO 300
      IF(SLOPE.LT..00001)GOTO 300
      FACT=1.486/XN*SQRT(SLOPE)
      YMIN=R*YMINK+R1*YMINJ
      YMAX=YMIN+100.
      DO 200 L=1,17
      YTEST=.5*(YMIN+YMAX)
      DY=YTEST-YFL
      YJ=YMINJ+DY
      YK=YMINK+DY
      CALL PROF3(J,YJ,AJ,V,PJ,DHJ,TWJ,XR,FR,1.,NT,ERROR)
      CALL PROF3(K,YK,AK,V,PK,DHK,TWK,XR,FR,1.,NT,ERROR)
      A=R*AK+R1*AJ
      P=R*PK+R1*PJ
      RH=A/P
      QTEST=FACT*A*RH**.6667
      IF(QTEST-Q)150,220,180
150   YMIN=YTEST
      GOTO 200
180   YMAX=YTEST
200   CONTINUE
220   YN=YTEST
300   CONTINUE
C
      RETURN
      END
C
C
C -------------------------------------------------------------------
      SUBROUTINE PROF5(NT)
C -------------------------------------------------------------------
      COMMON/BLK 3/TOL,Q,N,DIST(50),NE
C
      WRITE(NT,10)
10    FORMAT(//,35X,'(AES):',/,30X,'Irregular Channel',/,
     C 30X,'Subcritical Flow',/,34X,'Analysis',/,5X,
     C 'Study Name',30('_'),5X,'Page Number',15('_'),//)
      WRITE(NT,20)
20    FORMAT(5X,'The following study is based on the well known',/,
     C 5X,'STANDARD STEP METHOD to analyse subcritical flow in an',/,
     C 5X,'irregular channel. Energy-head losses and corresponding',/,
     C 5X,'notation used in the program are as follows:',/,
     C 5X,'    FRICTION LOSSES:   n = Mannings friction parameter',/,
     C 5X,'    EDDY LOSSES:       e = eddy loss coefficient',/,
```

```
     C  5X,'    KINETIC ENERGY',/,
     C  5X,'    CORRECTION FACTOR: a = correction factor')
        WRITE(NT,30)
30      FORMAT(/,5X,'The PROGRAM determines subcritical flow water',/,
     C  5X,'surface elevations by balancing the classical energy',/,
     C  5X,'equation between specified channel locations. All',/,
     C  5X,'geometric and parameter information is averaged between',/,
     C  5X,'defined channel cross-sections by straight-line',/,
     C  5X,'interpolation.')
        WRITE(NT,40)
40      FORMAT(/,5X,'THE ONLY LOSSES INCLUDED ARE FRICTION AND',/,
     C  5X,'EDDY LOSSES. The analysis formulation and presentation',/,
     C  5X,'of results follow the development given in',/,
     C  5X,'"OPEN CHANNEL HYDRAULICS", by Chow(1959).',//,
     C  5X,'The PROGRAM will default to a minimum flow-depth',/,
     C  5X,'of CRITICAL DEPTH, where the Froude number (Fr)',/,
     C  5X,'equals one (1). Therefore, supercritical water ')
        WRITE(NT,50)
50      FORMAT(5X,'surface information is not computed in this program.'
     C  ,/,5X,'In reaches of supercritical flow, critical depth is',/,
     C  5X,'assumed as a minimum flow depth which exceeds the actual',/,
     C  5X,'supercritical flowdepth in the channel.')
        WRITE(NT,60)Q,N,NE
60      FORMAT(/,5X,'For this study, the following information is used:'
     C  ,/,10X,'Channel flow(cfs) = ',F8.1,/,10X,
     C  'Number of channel cross-sections = ',I2,/,10X,
     C  'Number of Energy-Balance locations = ',I2)
        WRITE(NT,70)
70      FORMAT(//,5X,'Special notation given in the computer results',
     C  /,5X,'are as follows:',/,
     C  7X,'(1)FLOOD....this word appears in the first column',/,
     C  7X,'            whenever the estimated flowdepth exceeds either',/,
     C  7X,'            bank of the channel section.',/,
     C  7X,'(2)STEEP....this word appears in the last column',/,
     C  7X,'            whenever the channel Critical Depth exceeds',/,
     C  7X,'            the Normal Depth. Because this program')
        WRITE(NT,80)
80      FORMAT(7X,'          is intended only for subcritical flow,',/,
     C  7X,'            supercritical flow is modeled with flowdepths',/,
     C  7X,'            equal to or greater than Critical depth.',/,
     C  7X,'            This simplification results in conservative',/,
     C  7X,'            channel flowdepth estimations. However, supercritical',/,
     C  7X,'            flow effects including hydraulic jumps are',/,
     C  7X,'            NOT MODELED ACCURATELY.')
        WRITE(NT,90)
90      FORMAT(7X,'(3)SECTION..this word appears in the second column',/,
     C  7X,'            whenever the energy-balance channel location',/,
     C  7X,'            occurs at one of the defined channel cross-sections.')
C
        RETURN
        END
C
C
C ------------------------------------------------------------------
        SUBROUTINE PROF6(NT)
C ------------------------------------------------------------------
C
C
        COMMON/BLK 1/X(20,20),Y(20,20),NFL(20),M(20)
        COMMON/BLK 2/DISTS(20),ALPHA(20),XMAN(20),EDDY(20)
```

```
      COMMON/BLK 3/TOL,QQ,N,DIST(50),NE
C
C
C.......OUTPUT CROSS-SECTIONS
C
      IFF=12
C
      WRITE(NT,10)IFF
10    FORMAT(1X,A1,/)
      WRITE(NT,5)
5     FORMAT(76('*'))
C
      WRITE(NT,161)
      WRITE(NT,5)
C
      WRITE(NT,7)
7     FORMAT(5X,'CROSS-SECTION INFORMATION:')
      WRITE(NT,6)
C
C
      DO 50 K=1,N
      WRITE(NT,20)K,XMAN(K),ALPHA(K),EDDY(K),DISTS(K)
20    FORMAT(5X,'INFORMATION FOR CROSS-SECTION NUMBER:',I3,/,
     C 5X,'MANNINGS FRICTION FACTOR = ',F6.5,/,
     C 5X,'KINETIC ENERGY CORRECTION FACTOR = ',F5.3,/,
     C 5X,'EDDY LOSS FACTOR = ',F5.3,/,
     C 5X,'DISTANCE(ft.) TO CROSS-SECTION #1 = ',F8.2,/)
      MM=M(K)
      WRITE(NT,25)
25    FORMAT(5X,'NODAL POINT COORDINATE INFORMATION:',/,
     C 5X,'NODE NO.',5X,'X(ft.)',4X,'Y(elev.)')
      DO 30 KK=1,MM
      WRITE(NT,31)KK,X(K,KK),Y(K,KK)
31    FORMAT(7X,I3,7X,F7.2,5X,F7.2)
30    CONTINUE
      WRITE(NT,6)
6     FORMAT(76('='))
50    CONTINUE
      WRITE(NT,163)
163   FORMAT(//)
      WRITE(NT,5)
      WRITE(NT,60)
60    FORMAT(5X,'USER-SPECIFIED ENERGY-BALANCE CHANNEL LOCATIONS:'
     C ,/,10X,'ENERGY BALANCE        DISTANCE TO',/,
     C 10X,   'LOCATION NUMBER     CROSS-SECTION #1')
      WRITE(NT,61)(I,DIST(I),I=1,NE)
61    FORMAT(18X,I2,13X,F8.2)
      WRITE(NT,5)
      CONTINUE
C
C  FORMATS
C
161   FORMAT(16X,'IRREGULAR CHANNEL SUBCRITICAL FLOW ANALYSIS')
162   FORMAT(76('-'))
C
      RETURN
      END
```

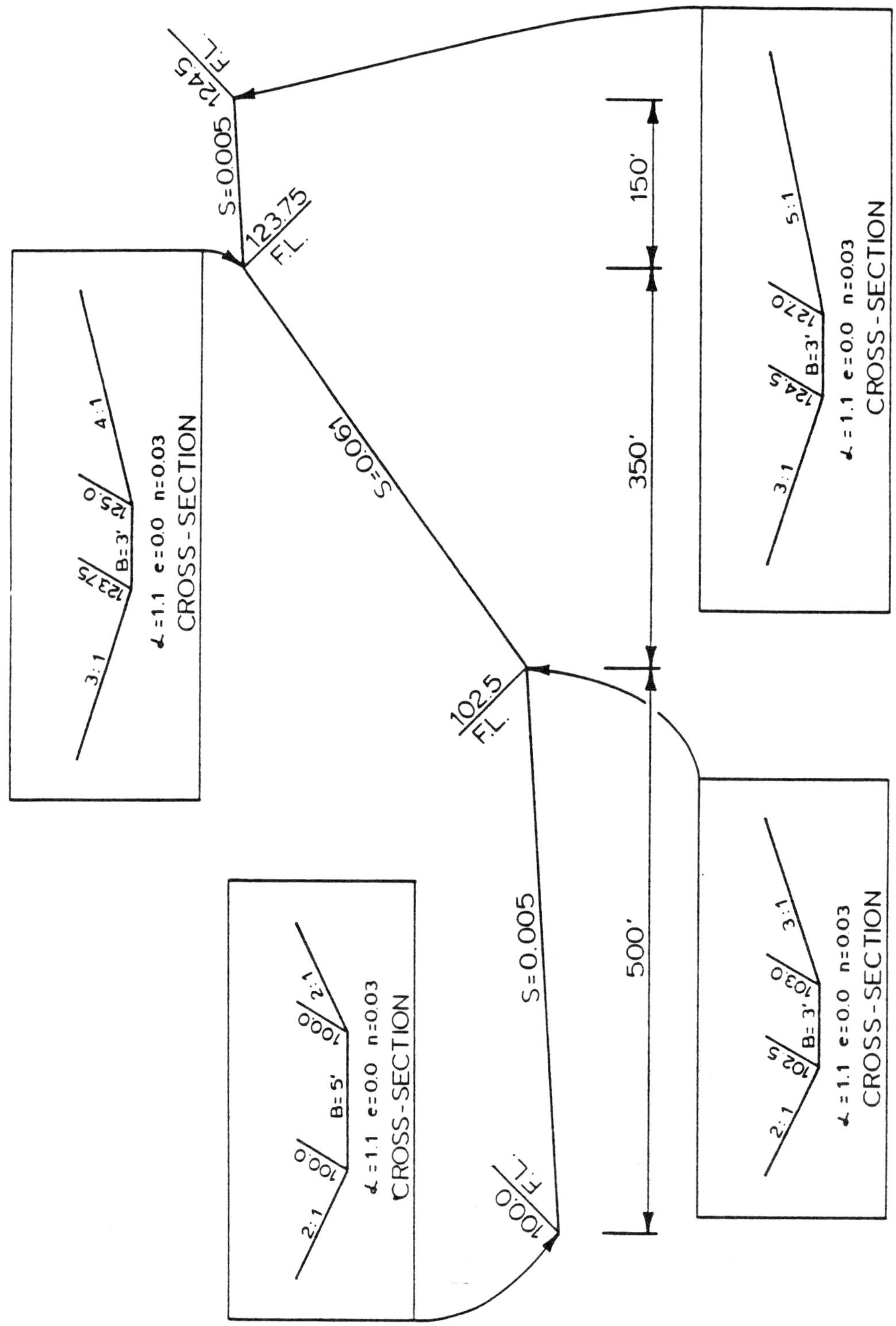

EXAMPLE PROBLEM
IRREGULAR CHANNEL PROFILE AND CROSS-SECTIONS

(AES):
Irregular Channel
Subcritical Flow
Analysis

Study Name______________________________ Page Number______________

The following study is based on the well known STANDARD STEP METHOD to analyse subcritical flow in an irregular channel. Energy-head losses and corresponding notation used in the program are as follows:

```
   FRICTION LOSSES:    n = Mannings friction parameter
   EDDY LOSSES:        e = eddy loss coefficient
   KINETIC ENERGY
   CORRECTION FACTOR: a = correction factor
```

The PROGRAM determines subcritical flow water surface elevations by balancing the classical energy equation between specified channel locations. All geometric and parameter information is averaged between defined channel cross-sections by straight-line interpolation.

THE ONLY LOSSES INCLUDED ARE FRICTION AND EDDY LOSSES. The analysis formulation and presentation of results follow the development given in "OPEN CHANNEL HYDRAULICS", by Chow(1959).

The PROGRAM will default to a minimum flow-depth of CRITICAL DEPTH, where the Froude number (Fr) equals one (1). Therefore, supercritical water surface information is not computed in this program. In reaches of supercritical flow, critical depth is assumed as a minimum flow depth which exceeds the actual supercritical flowdepth in the channel.

For this study, the following information is used:

```
     Channel flow(cfs) =   1000.0
     Number of channel cross-sections =  4
     Number of Energy-Balance locations = 11
```

Special notation given in the computer results are as follows:

(1)FLOOD....this word appears in the first column whenever the estimated flowdepth exceeds either bank of the channel section.

(2)STEEP....this word appears in the last column whenever the channel Critical Depth exceeds the Normal Depth. Because this program is intended only for subcritical flow, supercritical flow is modeled with flowdepths equal to or greater than Critical depth. This simplification results in conservative channel flowdepth estimations. However, supercritical flow effects including hydraulic jumps are NOT MODELED ACCURATELY.

(3)SECTION..this word appears in the second column whenever the energy-balance channel location occurs at one of the defined channel cross-sections.

```
**************************************************************************
         CROSS-SECTION INFORMATION:
**************************************************************************
     INFORMATION FOR CROSS-SECTION NUMBER:  1
     MANNINGS FRICTION FACTOR = .03000
     KINETIC ENERGY CORRECTION FACTOR = 1.100
     EDDY LOSS FACTOR =  .000
     DISTANCE(ft.) TO CROSS-SECTION #1 =      .00

     NODAL POINT COORDINATE INFORMATION:
     NODE NO.     X(ft.)    Y(elev.)
         1           .00      110.00
         2         20.00      100.00
         3         25.00      100.00
         4         45.00      110.00
==========================================================================
     INFORMATION FOR CROSS-SECTION NUMBER:  2
     MANNINGS FRICTION FACTOR = .03000
     KINETIC ENERGY CORRECTION FACTOR = 1.100
     EDDY LOSS FACTOR =  .000
     DISTANCE(ft.) TO CROSS-SECTION #1 =   500.00

     NODAL POINT COORDINATE INFORMATION:
     NODE NO.     X(ft.)    Y(elev.)
         1           .00      112.00
         2         19.00      102.50
         3         22.00      103.00
         4         46.00      111.00
==========================================================================
     INFORMATION FOR CROSS-SECTION NUMBER:  3
     MANNINGS FRICTION FACTOR = .03000
     KINETIC ENERGY CORRECTION FACTOR = 1.100
     EDDY LOSS FACTOR =  .000
     DISTANCE(ft.) TO CROSS-SECTION #1 =   850.00

     NODAL POINT COORDINATE INFORMATION:
     NODE NO.     X(ft.)    Y(elev.)
         1           .00      135.10
         2         34.00      123.75
         3         37.00      125.00
         4         77.00      135.00
==========================================================================
     INFORMATION FOR CROSS-SECTION NUMBER:  4
     MANNINGS FRICTION FACTOR = .03000
     KINETIC ENERGY CORRECTION FACTOR = 1.100
     EDDY LOSS FACTOR =  .000
     DISTANCE(ft.) TO CROSS-SECTION #1 =  1000.00

     NODAL POINT COORDINATE INFORMATION:
     NODE NO.     X(ft.)    Y(elev.)
         1           .00      136.00
         2         34.50      124.50
         3         37.50      127.00
         4         82.50      136.00
==========================================================================
**************************************************************************
     USER-SPECIFIED ENERGY-BALANCE CHANNEL LOCATIONS:
          ENERGY BALANCE          DISTANCE TO
          LOCATION NUMBER       CROSS-SECTION #1
                  1                  100.00
                  2                  200.00
                  3                  300.00
```

```
                  4              400.00
                  5              500.00
                  6              600.00
                  7              700.00
                  8              800.00
                  9              850.00
                 10              900.00
                 11             1000.00
*********************************************************************************
```

*** (AES): IRREGULAR CHANNEL SUBCRITICAL FLOW MODEL ***

Standard Step Method irregular channel analysis. Based on development in "OPEN CHANNEL HYDRAULICS",CHOW(1959)

STUDY NAME: Channel Flow = 1000.00 cfs PAGE NUMBER:

LENGTH from CONTROL	WATER SURFACE (elev.)	FLOW DEPTH (ft)	FLOW AREA (ft*ft)	FLOW V (fps)	2 aV /2g (ft)	TOTAL HEAD (ft)	HYDR RADIUS (ft)	FRICTION SLOPE Sf	AVERAGE REACH Sf	REACH LENGTH (ft)	LOSS Hf (ft)	EDDY LOSS (ft)	TOTAL HEAD (ft)	Fr
.0	105.91	5.91	99.4	10.06	1.729 a=1.10	107.639	3.16	.008885 n= .0300					107.639	1.00 GIVEN
100.0	107.21	6.71	123.5	8.10	1.121 a=1.10	108.335	3.47	.005061 n= .0300	.006973	100.0	.697	.00 e= .00	108.336	.77
200.0	107.73	6.73	123.6	8.09	1.119 a=1.10	108.845	3.42	.005155 n= .0300	.005108	100.0	.511	.00 e= .00	108.846	.78
300.0	108.26	6.76	124.4	8.04	1.105 a=1.10	109.362	3.38	.005177 n= .0300	.005166	100.0	.517	.00 e= .00	109.362	.78
400.0	108.79	6.79	125.4	7.97	1.087 a=1.10	109.880	3.34	.005171 n= .0300	.005174	100.0	.517	.00 e= .00	109.879	.78
500.0	109.33 SECTION	6.83	126.5	7.91	1.068 a=1.10	110.398	3.30	.005158 n= .0300	.005164	100.0	.516	.00 e= .00	110.397	.78
600.0	114.63	6.06	105.4	9.49	1.538 a=1.10	116.170	2.89	.008861 n= .0300	.007009	100.0	.701	.00 e= .00	111.099	*1.00 STEEP
700.0	120.62	5.98	107.4	9.31	1.482 a=1.10	122.106	2.81	.008875 n= .0300	.008868	100.0	.887	.00 e= .00	117.057	*1.00 STEEP
800.0	126.62	5.91	109.3	9.15	1.431 a=1.10	128.054	2.73	.008892 n= .0300	.008884	100.0	.888	.00 e= .00	122.994	*1.00 STEEP
850.0	129.63 SECTION	5.88	110.3	9.06	1.404 a=1.10	131.032	2.70	.008866 n= .0300	.008879	50.0	.444	.00 e= .00	128.498	*1.00 STEEP
900.0	130.24	6.24	119.3	8.38	1.200 a=1.10	131.438	2.76	.007355 n= .0300	.008110	50.0	.406	.00 e= .00	131.437	.91
1000.0	130.98 SECTION	6.48	118.4	8.45	1.220 a=1.10	132.202	2.65	.007902 n= .0300	.007628	100.0	.763	.00 e= .00	132.201	.93

PROGRAM 19: DATA ENTRY

```
---DATA ENTRY FOR IRREGULAR CHANNEL HYDRAULICS---PAGE 1


Enter TOTAL NUMBER of subchannels in irregular
 channel......................................... ===>     "N"
:ALLOWABLE VALUES ARE [1] TO [10    ]

Enter irregular channel SLOPE(FEET/FEET)............ ===>     "S"
:ALLOWABLE VALUES ARE [.00001] TO [.25    ]

Enter TOTAL irregular channel FLOW(CFS)............. ===>     "QQ"
:ALLOWABLE VALUES ARE [1      ] TO [999999]

Enter allowable ERROR(FEET) in normal depth
determination....................................... ===>     "TOL"
:ALLOWABLE VALUES ARE [.01    ] TO [.50    ]



---------------------------------------------------------------------------
TYPE: EXIT to leave program ; TOP  to go to top of page ; HELP for Help
      MAIN to go to main menu
```

```
---DATA ENTRY FOR IRREGULAR CHANNEL HYDRAULICS---

SUBCHANNEL DATA ENTRY GROUP  1 OF  1:

Enter TOTAL number of cross-section nodal points
to describe subchannel............................... ===>     "M(I)"
:ALLOWABLE VALUES ARE [2] TO [20     ]
(NOTE: INCLUDE BOTH END POINTS IN TOTAL NUMBER)

Enter Mannings friction factor....................... ===>     "XMAN(I)"
:ALLOWABLE VALUES ARE [.008   ] TO [.99    ]



------------------------------------------------------------------------------
    TYPE: EXIT to leave program ; TOP  to go to top of page
                                ; BACK to go back one page
```

```
---DATA ENTRY FOR IRREGULAR CHANNEL HYDRAULICS---
             CROSS-SECTION   1 NODAL POINT  2 OF  3:

Enter (X) coordinate................................. ===>     "X(I,J)"
:ALLOWABLE VALUES ARE [       .01] TO [9999      ]

Enter (Y) elevation.................................. ===>     "Y(I,J)"
:ALLOWABLE VALUES ARE [-9999     ] TO [9999      ]

NODE  (X) COORDINATE      (Y) ELEVATION
  1       [    .00]          [  20.00]
          [       ]          [       ]
          [       ]          [       ]
          [       ]          [       ]
          [       ]          [       ]
          [       ]          [       ]
          [       ]          [       ]
          [       ]          [       ]
          [       ]          [       ]
          [       ]          [       ]

------------------------------------------------------------------------------
    TYPE: EXIT to leave program ; TOP  to go to top of page
                                ; BACK to go back one entry
```

PROGRAM 19

```
C ----------------------------------------------------------------
        SUBROUTINE DEPTH
C ----------------------------------------------------------------
C
C  DATA ENTRY ROUTINE FOR DEPTH BATCH SYSTEM
C
        COMMON/BLK1/X(20,20),Y(20,20),NFL(20),M(20)
        COMMON/BLK2/DISTS(20),ALPHA(20),XMAN(20),EDDY(20)
        COMMON/BLK3/TOL,QQ,N,DIST(50),NE
C
C
        NUT=6
C
C..READ DATA INPUT
        READ FREE(5) N,S,QQ,TOL
        DO 10 I=1,N
        READ FREE(5) M(I),XMAN(I)
        MM=M(I)
        DO 20 J=1,MM
        READ FREE(5) X(I,J),Y(I,J)
20      CONTINUE
10      CONTINUE
C
C
C..END OF INPUT
C
        CALL NORM2A(X,Y,XMAN,M,S,TOL,NUT,QQ,N)
C
C
        RETURN
        END
C
C
C ----------------------------------------------------------------
        SUBROUTINE NORM2A(X,Y,XMAN,M,S,TOL,NT,QQ,N)
C ----------------------------------------------------------------
C
C
        DIMENSION X(20,20),Y(20,20),XMAN(20),M(20),A(10),TW(10),Q(10)
        DIMENSION V(10),FR(10),HD(10)
        YLEFT=Y(1,1)
        YDEPTH=20000.
        DO 40 I=1,N
        DO 41 J=1,M(I)
        IF(Y(I,J).LT.YDEPTH)YDEPTH=Y(I,J)
        YRIGHT=Y(I,J)
41      CONTINUE
40      CONTINUE
50      YDEPTH=YDEPTH+TOL
        QTOT=0.
        DO 100 I=1,N
        L=I
        CALL NORM3A(X,Y,YDEPTH,XMAN,L,M,A,V,Q,S,TW)
        QTOT=QTOT+Q(I)
100     CONTINUE
        IF(QTOT-QQ)50,110,110
110     DO 300 I=1,N
        IF(ABS(TW(I)).LT..0001)GOTO 300
        HD(I)=A(I)/TW(I)
```

```
          FR(I)=0.
          FR(I)=V(I)/((32.2*HD(I))**.5)
300       CONTINUE
500       FORMAT(76('-'))
602       FORMAT(76('*'))
C
          WRITE(NT,500)
C
          DO 550 I=1,N
          WRITE(NT,504)I
504       FORMAT(13X,'* ENTERED INFORMATION FOR SUBCHANNEL NUMBER',I3,' :')
          WRITE(NT,505)
505       FORMAT(13X,'NODE NUMBER',6X,'"X" COORDINATE',5X,'"Y" ',
     .'COORDINATE',/)
          DO 510 J=1,M(I)
          WRITE(NT,506)J,X(I,J),Y(I,J)
506       FORMAT(17X,I2,13X,F9.2,10X,F9.2)
510       CONTINUE
          WRITE(NT,507)S,XMAN(I)
507       FORMAT(13X,'SUBCHANNEL SLOPE(FEET/FEET) = ',F8.6,/,
     C    13X,'SUBCHANNEL MANNINGS FRICTION FACTOR = ',F8.6,/)
          WRITE(NT,508)
508       FORMAT(76('.'))
          WRITE(NT,509)Q(I),A(I),V(I),FR(I),TW(I),HD(I)
509       FORMAT(10X,'SUBCHANNEL FLOW(CFS) = ',F9.1,/,
     C    10X,'SUBCHANNEL FLOW AREA(SQUARE FEET) = ',F10.2,/,
     C    10X,'SUBCHANNEL FLOW VELOCITY(FEET/SEC.) = ',F8.3,/,
     C    10X,'SUBCHANNEL FROUDE NUMBER = ',F6.3,/,
     C    10X,'SUBCHANNEL FLOW TOP-WIDTH(FEET) = ',F10.2,/,
     C    10X,'SUBCHANNEL HYDRAULIC DEPTH(FEET) = ',F7.2,/)
          WRITE(NT,500)
          WRITE(NT,500)
550       CONTINUE
          WRITE(NT,560)QQ,QTOT,YDEPTH
560    FORMAT(10X,'TOTAL IRREGULAR CHANNEL FLOW(CFS) WANTED = ',F10.2,/,
     C    10X,'COMPUTED IRREGULAR CHANNEL FLOW(CFS) = ',F10.2,//,
     C    10X,'ESTIMATED IRREGULAR CHANNEL NORMAL DEPTH WATER SURFACE',/,
     C    10X,'ELEVATION...........................',F10.2,//)
          IF((YDEPTH.GT.YLEFT).OR.(YDEPTH.GT.YRIGHT))GOTO 595
          WRITE(NT,580)
580       FORMAT(15X,'NOTE: WATER SURFACE IS BELOW EXTREME',/,
     C    21X,'LEFT AND RIGHT BANK ELEVATIONS.',/)
          GOTO 600
595       WRITE(NT,585)
585       FORMAT(15X,'NOTE: WATER SURFACE IS ABOVE LEFT OR RIGHT',/,
     C    21X,'BANK ELEVATIONS.',/)
600       WRITE(NT,500)
C
          RETURN
          END
C
C
C   ------------------------------------------------------------------
          SUBROUTINE NORM3A(X,Y,YDEPTH,XMAN,L,M,A,V,Q,S,TW)
C   ------------------------------------------------------------------
C
C
          DIMENSION X(20,20),Y(20,20),XMAN(20),M(20),A(10)
          DIMENSION V(10),Q(10),P(10),TW(10)
          HYPOT(B,C)=((B*B)+(C*C))**.5
```

```
      I=L
      P(I)=0.
      V(I)=0.
      Q(I)=0.
      TW(I)=0.
      A(I)=0.
      KK=M(I)-1
      DO 200 J=1,KK
      IF((YDEPTH.LE.Y(I,J)).AND.(YDEPTH.LE.Y(I,(J+1))))GOTO 200
      XDIF=X(I,(J+1))-X(I,J)
      YDIF=Y(I,(J+1))-Y(I,J)
      IF(YDIF)150,150,160
150   YDIF=-YDIF
      YUP=Y(I,J)
      YLOW=Y(I,(J+1))
      GOTO 170
160   YUP=Y(I,(J+1))
      YLOW=Y(I,J)
170   TEMP=YUP-YDEPTH
      IF(TEMP.LE.0.)GOTO 155
      GOTO 165
155   AA=((YDEPTH-YUP)*XDIF)+(.5*XDIF*YDIF)
      PP=HYPOT(XDIF,YDIF)
      TWTW=XDIF
      GOTO 198
165   IF(ABS(YDIF).LT..0001)GOTO 155
      YDEEP=YDEPTH-YLOW
      TWTW=XDIF*(YDEEP/YDIF)
      AA=.5*YDEEP*TWTW
      PP=(YDEEP/YDIF)*HYPOT(XDIF,YDIF)
198   A(I)=A(I)+AA
      P(I)=P(I)+PP
      TW(I)=TW(I)+TWTW
200   CONTINUE
      IF(P(I).LE..0001)GOTO 180
      R=(A(I)/P(I))**.66667
      V(I)=1.486*R*(S**.5)/XMAN(I)
      Q(I)=V(I)*A(I)
180   CONTINUE
C
      RETURN
      END
```

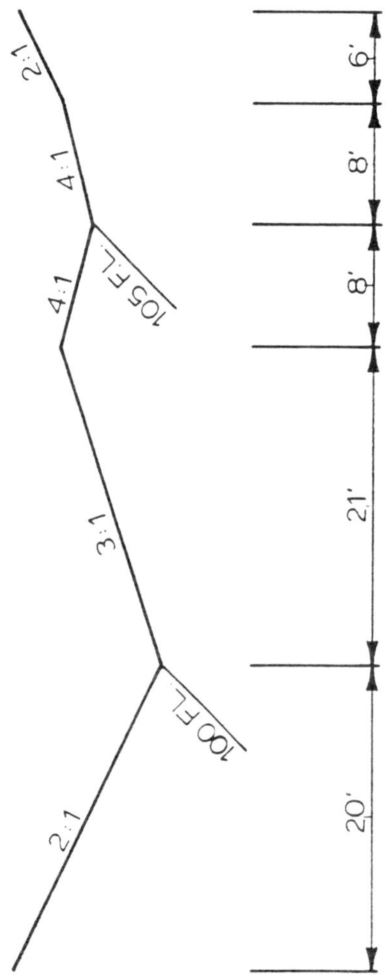

EXAMPLE PROBLEM – IRREGULAR CHANNEL CROSS-SECTION

```
**********DESCRIPTION OF RESULTS********************************************
*                                                                          *
*                           EXAMPLE PROBLEM                                *
*                                                                          *
****************************************************************************


<<<<<<<<<<<<<<<<<<<<<<<<<<<<<<<<<<<<<<>>>>>>>>>>>>>>>>>>>>>>>>>>>>>>>>>>>>>>

                  Advanced Engineering Software [AES]
                            SERIAL No. I0008
                  REV. 1.0      RELEASE DATE: 7/02/81

<<<<<<<<<<<<<<<<<<<<<<<<<<<<<<<<<<<<<<>>>>>>>>>>>>>>>>>>>>>>>>>>>>>>>>>>>>>>

----------------------------------------------------------------------------
            * ENTERED INFORMATION FOR SUBCHANNEL NUMBER  1 :
            NODE NUMBER      "X" COORDINATE     "Y" COORDINATE

                 1                   .00             110.00
                 2                 20.00             100.00
                 3                 41.00             107.00
            SUBCHANNEL SLOPE(FEET/FEET) =  .005000
            SUBCHANNEL MANNINGS FRICTION FACTOR =  .030000

............................................................................
         SUBCHANNEL FLOW(CFS) =      714.6
         SUBCHANNEL FLOW AREA(SQUARE FEET) =       99.75
         SUBCHANNEL FLOW VELOCITY(FEET/SEC.) =     7.164
         SUBCHANNEL FROUDE NUMBER =    .710
         SUBCHANNEL FLOW TOP-WIDTH(FEET) =       31.58
         SUBCHANNEL HYDRAULIC DEPTH(FEET) =     3.16

----------------------------------------------------------------------------
----------------------------------------------------------------------------
            * ENTERED INFORMATION FOR SUBCHANNEL NUMBER  2 :
            NODE NUMBER      "X" COORDINATE     "Y" COORDINATE

                 1                 41.00             107.00
                 2                 49.00             105.00
                 3                 57.00             107.00
                 4                 63.00             110.00
            SUBCHANNEL SLOPE(FEET/FEET) =  .005000
            SUBCHANNEL MANNINGS FRICTION FACTOR =  .015000
............................................................................
         SUBCHANNEL FLOW(CFS) =       36.0
         SUBCHANNEL FLOW AREA(SQUARE FEET) =        6.93
         SUBCHANNEL FLOW VELOCITY(FEET/SEC.) =     5.195
         SUBCHANNEL FROUDE NUMBER = 1.128
         SUBCHANNEL FLOW TOP-WIDTH(FEET) =       10.53
         SUBCHANNEL HYDRAULIC DEPTH(FEET) =      .66

----------------------------------------------------------------------------
----------------------------------------------------------------------------
         TOTAL IRREGULAR CHANNEL FLOW(CFS) WANTED =      750.00
         COMPUTED IRREGULAR CHANNEL FLOW(CFS) =      750.60

         ESTIMATED IRREGULAR CHANNEL NORMAL DEPTH WATER SURFACE
         ELEVATION.............................    106.32


              NOTE: WATER SURFACE IS BELOW EXTREME
                    LEFT AND RIGHT BANK ELEVATIONS.

----------------------------------------------------------------------------
```

PROGRAM 20: DATA ENTRY

```
---DATA ENTRY FOR RETARDING BASIN ROUTING---PAGE 1

RESERVOIR INFLOW HYDROGRAPH INFORMATION:

Enter constant hydrograph interval time
unit(MINUTES)...................................... ===>    "DELT"
:ALLOWABLE VALUES ARE [1] TO [120   ]


Enter total number of inflow hydrograph intervals... ===>    "NUMBER"
:ALLOWABLE VALUES ARE [5] TO [100   ]









-------------------------------------------------------------------------------
    TYPE: EXIT to leave program ; TOP  to go to top of page
```

```
---DATA ENTRY FOR RETARDING BASIN ROUTING---PAGE 2

INFLOW-HYDROGRAPH FLOW DATA ENTRY   1 OF  10

Enter hydrograph interval flow(CFS)................. ===>    "H(I)"
:ALLOWABLE VALUES ARE [0] TO [1000000   ]

         INTERVAL    HYDROGRAPH
          NUMBER        FLOW

          [  ]     [          ]
          [  ]     [          ]
          [  ]     [          ]
          [  ]     [          ]
          [  ]     [          ]
          [  ]     [          ]
          [  ]     [          ]
          [  ]     [          ]
          [  ]     [          ]
          [  ]     [          ]

-------------------------------------------------------------------------------
    TYPE: EXIT to leave program ; TOP  to go to top of page
                                ; BACK to go back one page
```

```
---DATA ENTRY FOR RETARDING BASIN ROUTING---PAGE 3


The model requires DEPTH-VERSES-STORAGE and DISCHARGE
information.  On the next page, Basin Depth, Storage, and
Outflow are entered as a group.  For DEPTH-VS.-STORAGE
data entry group # 1 only:
     Bottom of basin assumed as depth(FEET) = 0
     Basin outflow at depth = 0 FEET assumed as 0 CFS
     Basin storage at depth = 0 FEET assumed as 0 ACRE-FEET

Enter total number of tabulation entry groups....... ===>    "NUMD"
:ALLOWABLE VALUES ARE [3] TO [20    ]






-------------------------------------------------------------------------------
    TYPE: EXIT to leave program ; TOP  to go to top of page
                                ; BACK to go back one page
```

```
---DATA ENTRY FOR RETARDING BASIN ROUTING---PAGE 4

DEPTH-VS.-STORAGE AND DEPTH-VS.-DISCHARGE DATA ENTRY GROUP  2 OF  3:

Enter basin depth(FEET)............................. ===>    "D(I)"
:ALLOWABLE VALUES ARE [      .050] TO [200        ]

Enter basin storage(ACRE-FEET)...................... ===>    "S(I)"
:ALLOWABLE VALUES ARE [      .000] TO [1000000    ]

Enter basin outflow(CFS)............................ ===>    "O(I)"
:ALLOWABLE VALUES ARE [      .000] TO [1000000    ]

 DATA ENTRY    BASIN DEPTH       STORAGE       OUTFLOW
    GROUP         (FEET)       (ACRE-FEET)       (CFS)
      1       [        .000]  [        .000]  [        .000]
              [            ]  [            ]  [            ]
              [            ]  [            ]  [            ]
              [            ]  [            ]  [            ]
              [            ]  [            ]  [            ]

-------------------------------------------------------------------------------
    TYPE: EXIT to leave program ; TOP  to go to top of page
                                ; BACK to go back one page
```

```
---DATA ENTRY FOR RETARDING BASIN ROUTING---PAGE 5



Enter assumed initial depth(FEET) of water in basin
at BEGINNING of inflow hydrograph into basin........ ===>    "DO"
:ALLOWABLE VALUES ARE [0] TO      20.00 ]


Enter maximum number of outflow hydrograph
intervals desired................................... ===>    "LIMIT"
:ALLOWABLE VALUES ARE [10    ] TO [1000  ]








--------------------------------------------------------------------------------
    TYPE: EXIT to leave program ; TOP  to go to top of page
                                ; BACK to go back one page
```

PROGRAM 20

```
C  -----------------------------------------------------------------
         SUBROUTINE RETARD
C  -----------------------------------------------------------------
C
C  DATA ENTRY ROUTINE FOR RETARDING BASIN BATCH SYSTEM
C
         COMMON/NUT/NUT
         COMMON/BLK1A/H(300)
         COMMON/BLK2A/S(20)
         COMMON/BLK3A/O(20)
         COMMON/BLK4A/D(20)
         COMMON/BLK5A/A(20)
         COMMON/BLK6A/B(20)
C
C
         NUT=6
C
         OPEN 5,"RETRD.DAT"
         OPEN 6,"RETRD.ANS"
C
C
C..READ DATA INPUT
         READ FREE(5) DELT,NUMBER
         DO 10 I=1,NUMBER
         READ FREE(5) H(I)
10       CONTINUE
C
         READ FREE(5) NUMD
         DO 20 I=1,NUMD
         READ FREE(5) D(I)
         READ FREE(5) S(I)
         READ FREE(5) O(I)
20       CONTINUE
C
         READ FREE(5) D0,LIMIT
C
C..END OF INPUT
C
         CALL BASIN(NUMBER,NUMD,DELT,D0,NUT,LIMIT)
C
         RETURN
         END
C
C
C  -----------------------------------------------------------------
         SUBROUTINE BASIN(NUMBER,NUMD,DELT,D0,NUT,LIMIT)
C  -----------------------------------------------------------------
C
C  THIS SUBROUTINE ROUTS HYDROGRAPH THRU BASIN
C
         INTEGER BLANK,DOT,CROSS,DASH,LINE(33)
         COMMON/BLK1A/H(300)
         COMMON/BLK2A/S(20)
         COMMON/BLK3A/O(20)
         COMMON/BLK4A/D(20)
         COMMON/BLK5A/A(20)
         COMMON/BLK6A/B(20)
C
         DATA BLANK,DOT,CROSS,DASH,LINE/' ','.','I','O',33*' '/
```

```
C
C
C.......GET INITIAL VALUES
        CALL SEE(D0,1,D1,D2,I1,I2,NUT)
        O2=0.
        S2=0.
        S1=S(I1)+(D0-D1)/(D2-D1)*(S(I2)-S(I1))
        O1=O(I1)+(D0-D1)/(D2-D1)*(O(I2)-O(I1))
        WRITE(NUT,2)
        WRITE(NUT,10)D0,S1,O1
10      FORMAT(5X,'INITIAL BASIN DEPTH(FEET) = ',F7.2,/,
     C  5X,'INITIAL BASIN STORAGE(ACRE-FEET) = ',F10.2,/,
     C  5X,'INITIAL BASIN OUTFLOW(CFS) = ',F10.2)
        WRITE(NUT,2)
2       FORMAT(4('-------------------'))
C
C  CONSTRUCT S+-O/2DELT CURVES
C
C  USE A-VECTOR TO STORE {S}-{O}*DELT/2
C  USE B-VECTOR TO STORE {S}+{O}*DELT/2
        CON=60./43560.*DELT/2.
        DO 100 K=1,NUMD
        A(K)=S(K)-O(K)*CON
100     B(K)=S(K)+O(K)*CON
C
C  SET A(1)=B(1)
        A(1)=B(1)
        WRITE(NUT,2)
        WRITE(NUT,101)
101     FORMAT(5X,'BASIN STORAGE, OUTFLOW AND DEPTH ROUTING VALUES:',/,
     C  5X,' INTERVAL    {S-O*DT/2}    {S+O*DT/2}',/,5X,
     C  '  NUMBER    (ACRE-FEET) (ACRE-FEET)')
        WRITE(NUT,102)(K,A(K),B(K),K=1,NUMD)
102     FORMAT(5X,I7,1X,2F12.5)
        WRITE(NUT,103)
103    FORMAT(5X,'WHERE S=STORAGE(AF);O=OUTFLOW(AF/MIN.);DT=UNIT(MIN.)')
        WRITE(NUT,2)
        CON=CON*2.
        ATEMP=S1-O1*CON/2.
        K=0
        DEPTH1=D0
C
C  FIND XMAX
C
        XMAX=0.
        DO 198 K=1,NUMBER
        IF(H(K).GT.XMAX)XMAX=H(K)
198     CONTINUE
        F=32./XMAX
        T0=0.
        T1=XMAX/4.
        T2=2.*T1
        T3=3.*T1
        T4=4.*T1
        WRITE(NUT,776)
776     FORMAT(/////,10X,'*UNIT-HYDROGRAPH STORAGE-BASIN ROUTING*',/,5X,
     C  'NOTE: COMPUTED BASIN DEPTH, OUTFLOW, AND STORAGE QUANTITIES',/,
     C11X,'OCCUR AT THE GIVEN TIME. BASIN INFLOW VALUES REPRESENT THE',/
     C,11X,'AVERAGE INFLOW DURING THE RECENT HYDROGRAPH UNIT INTERVAL.'
     C  //,5X,'GRAPH NOTATION: "I"=MEAN UNIT INFLOW; "O"=OUTFLOW AT',
```

```
     C  ' GIVEN TIME')
        WRITE(NUT,2)
        WRITE(NUT,105)T0,T1,T2,T3,T4
105     FORMAT(6X,'TIME',3X,'INFLOW',2X,'OUTFLOW',4X,'STORAGE',/,
     C  5X,'(HOURS)',1X,'(CFS)',4X,'(CFS)',5X,' ACRE-FT',1X,F2.1,4F8.0)
        K=0
200     K=K+1
        IF(K.GT.LIMIT)GOTO 1000
        QQ=0.
        IF(K.LE.NUMBER)QQ=CON*H(K)
        IF(K.GT.NUMBER)H(K)=0.
        TEMP=QQ+ATEMP
        CALL SEE(TEMP,2,B1,B2,I1,I2,NUT)
        DEPTH2=D(I1)+(TEMP-B1)/(B2-B1)*(D(I2)-D(I1))
        S2=S(I1)+(DEPTH2-D(I1))/(D(I2)-D(I1))*(S(I2)-S(I1))
        O2=O(I1)+(DEPTH2-D(I1))/(D(I2)-D(I1))*(O(I2)-O(I1))
        ATEMP=S2-O2*CON/2.
        XK=K
        TIME=XK*DELT/60.
        TEST=F*H(K)
        TEST1=F*O2
        LINE(1)=DOT
        LINE(9)=DOT
        LINE(17)=DOT
        LINE(25)=DOT
        LINE(33)=DOT
        JJ=TEST+1
        LINE(JJ)=CROSS
        JJJ=TEST1+1
        LINE(JJJ)=DASH
        WRITE(NUT,777)TIME,H(K),O2,S2,LINE
777     FORMAT(5X,F7.2,2F9.1,F12.3,1X,33A1)
        LINE(JJ)=BLANK
        LINE(JJJ)=BLANK
        WRITE(NUT,778)DEPTH2
778     FORMAT(10X,'[BASIN DEPTH(FEET) = ',F6.1,']')
        DEPTH1=DEPTH2
        GOTO 200
1000    CONTINUE
C
        RETURN
        END
C
C
C  -------------------------------------------------------------
        SUBROUTINE SEE(X,MM,D1,D2,I1,I2,NUT)
C  -------------------------------------------------------------
C
C
        DIMENSION Y(20)
        COMMON/BLK4A/D(20)
        COMMON/BLK6A/B(20)
C
C
        IF(MM.EQ.2)GOTO 20
        DO 10 K=1,20
10      Y(K)=D(K)
        GOTO 30
20      DO 25 K=1,20
25      Y(K)=B(K)
```

```
30      CONTINUE
        DO 100 K=1,20
        IF(X.LT.Y(K))GOTO 200
100     CONTINUE
        WRITE(NUT,101)
101     FORMAT(5X,'NOTE: BASIN LIMITATIONS ARE EXCEEDED.',/,5X,
     C  'STORAGE, DEPTH, AND OUTFLOW INFORMATION ARE EXTRAPOLATIONS.')
        GOTO 1000
200     I1=K-1
        I2=K
        D1=Y(I1)
        D2=Y(I2)
1000    CONTINUE
C
        RETURN
        END
```

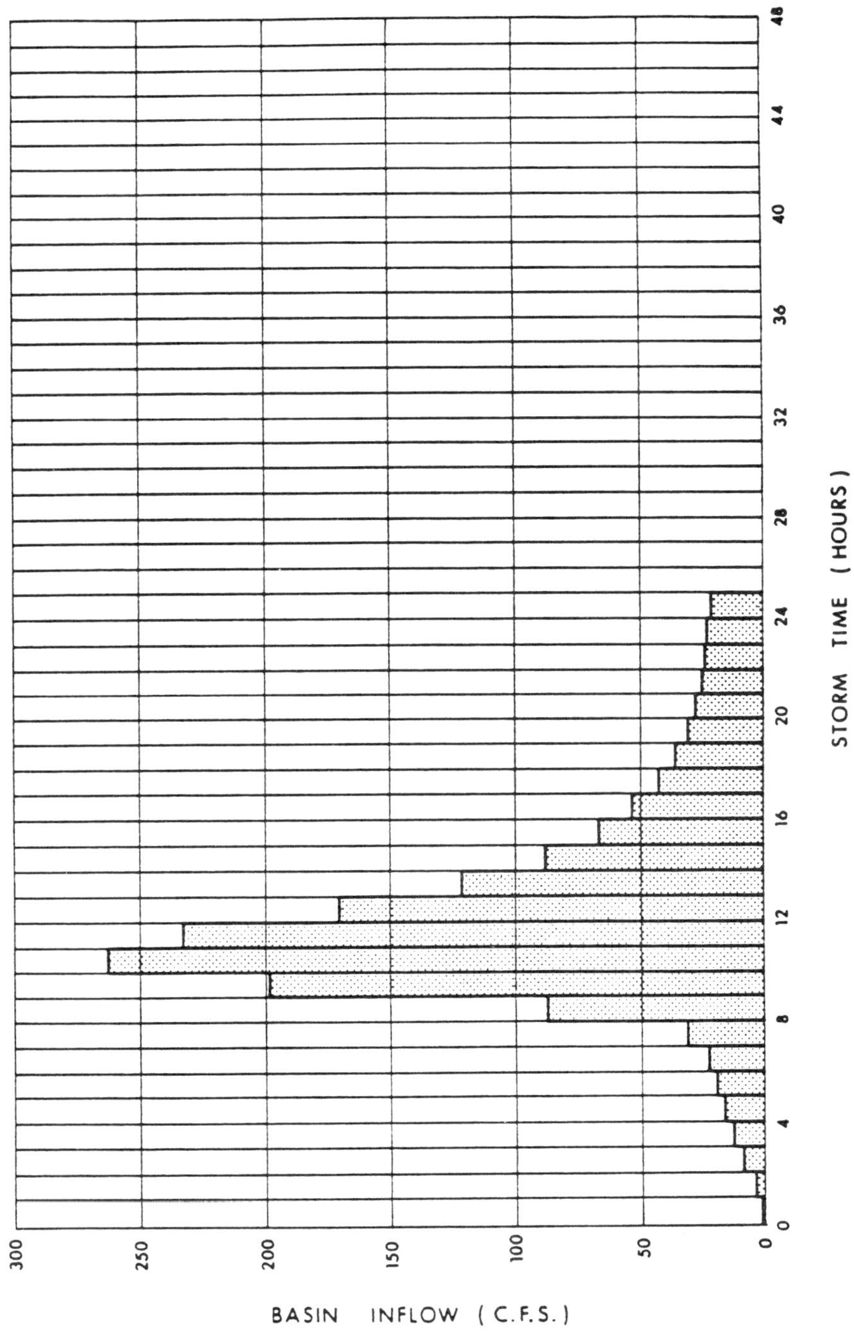

EXAMPLE PROBLEM – INFLOW HYDROGRAPH

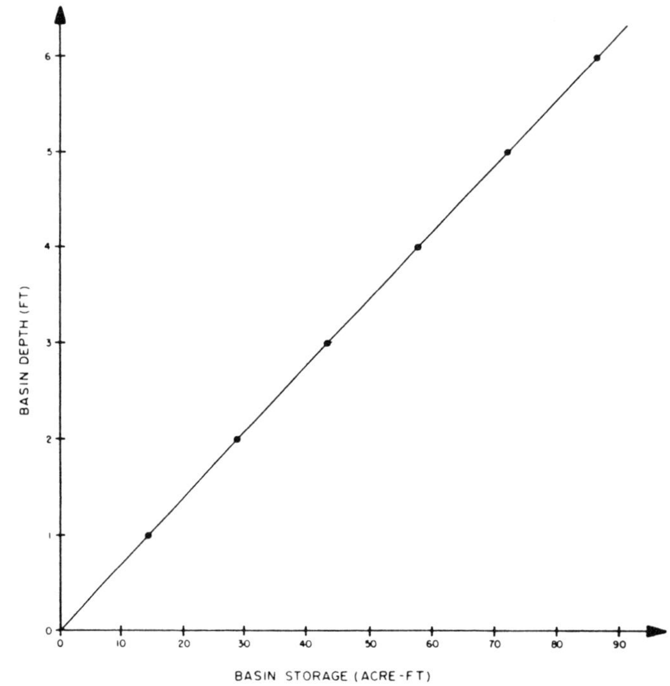

EXAMPLE PROBLEM
DETENTION BASIN STORAGE (AF) AS A FUNCTION OF DEPTH (FT)

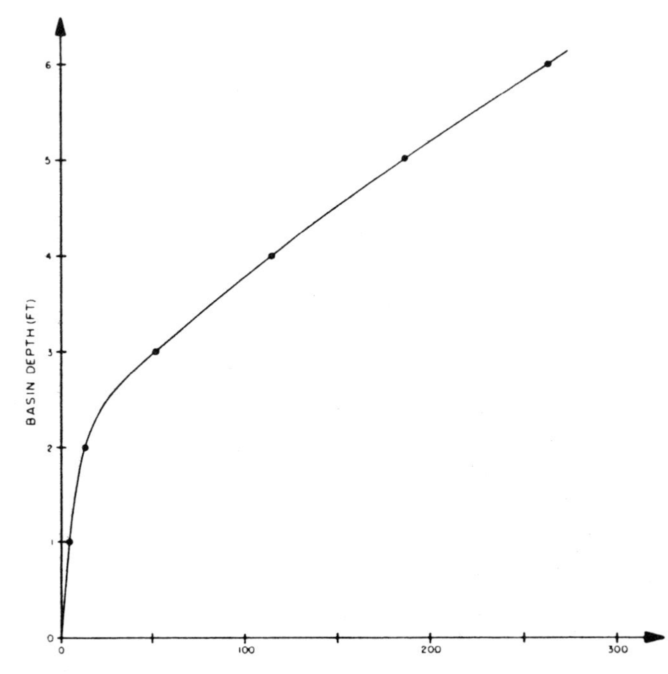

EXAMPLE PROBLEM
DETENTION BASIN OUTFLOW (CFS) AS A FUNCTION OF DEPTH (FT)

```
=============================================================================
    ENTERED INFORMATION:
-----------------------------------------------------------------------------
    TOTAL NUMBER OF INFLOW HYDROGRAPH INTERVALS =   25
    CONSTANT HYDROGRAPH TIME UNIT(MINUTES) =   60.000
    ASSUMED INITIAL DEPTH(FEET) IN STORAGE BASIN =      .00
-----------------------------------------------------------------------------
    ENTERED INFLOW HYDROGRAPH ORDINATES(CFS):
    *INTERVAL     FLOW   *INTERVAL     FLOW   *INTERVAL     FLOW   *
    * NUMBER     (CFS)   * NUMBER     (CFS)   * NUMBER     (CFS)   *
    *      1:       .70*       2:       3.50*       3:       8.10*
    *      4:     12.50*       5:      16.10*       6:      19.20*
    *      7:     22.30*       8:      31.30*       9:      87.90*
    *     10:    198.30*      11:     263.20*      12:     233.20*
    *     13:    170.50*      14:     121.30*      15:      88.30*
    *     16:     66.90*      17:      52.50*      18:      42.60*
    *     19:     35.30*      20:      30.40*      21:      27.30*
    *     22:     25.20*      23:      23.90*      24:      23.00*
    *     25:     20.30*
=============================================================================
    DEPTH-VS.-STORAGE AND DEPTH-VS.-DISCHARGE INFORMATION:
-----------------------------------------------------------------------------
    TOTAL NUMBER OF BASIN DEPTH INFORMATION ENTRIES =   7
-----------------------------------------------------------------------------
    *BASIN-DEPTH  STORAGE    OUTFLOW  **BASIN-DEPTH  STORAGE     OUTFLOW  *
    *   (FEET)  (ACRE-FEET)   (CFS)   **    (FEET)  (ACRE-FEET)   (CFS)   *
    *      .000        .000       .000**     1.000     14.400      4.240*
    *     2.000      28.800     12.000**     3.000     43.200     51.710*
    *     4.000      57.600    114.670**     5.000     72.000    186.810*
    *     6.000      86.400    263.190**
*****************************************************************************

<<<<<<<<<<<<<<<<<<<<<<<<<<<<<<<<<<<<<<>>>>>>>>>>>>>>>>>>>>>>>>>>>>>>>>>>>>>>>

                  Advanced Engineering Software [AES]
                          SERIAL No. E0008
                  REV. 1.0     RELEASE DATE: 6/18/81

<<<<<<<<<<<<<<<<<<<<<<<<<<<<<<<<<<<<<<>>>>>>>>>>>>>>>>>>>>>>>>>>>>>>>>>>>>>>>

*****************************************************************************
-----------------------------------------------------------------------------
    INITIAL BASIN DEPTH(FEET) =      .00
    INITIAL BASIN STORAGE(ACRE-FEET) =         .00
    INITIAL BASIN OUTFLOW(CFS) =        .00
-----------------------------------------------------------------------------
-----------------------------------------------------------------------------
    BASIN STORAGE, OUTFLOW AND DEPTH ROUTING VALUES:
```

```
 INTERVAL   {S-O*DT/2}    {S+O*DT/2}
  NUMBER   (ACRE-FEET) (ACRE-FEET)
      1       .00000       .00000
      2     14.22479     14.57521
      3     28.30412     29.29587
      4     41.06322     45.33676
      5     52.86157     62.33842
      6     64.28058     79.71941
      7     75.52437     97.27560
WHERE S=STORAGE(AF);O=OUTFLOW(AF/MIN.);DT=UNIT(MIN.)
------------------------------------------------------------------------------
```

```
     *UNIT-HYDROGRAPH STORAGE-BASIN ROUTING*
NOTE: COMPUTED BASIN DEPTH, OUTFLOW, AND STORAGE QUANTITIES
      OCCUR AT THE GIVEN TIME. BASIN INFLOW VALUES REPRESENT THE
      AVERAGE INFLOW DURING THE RECENT HYDROGRAPH UNIT INTERVAL.

GRAPH NOTATION: "I"=MEAN UNIT INFLOW; "O"=OUTFLOW AT GIVEN TIME
------------------------------------------------------------------------------
 TIME    INFLOW  OUTFLOW     STORAGE
(HOURS) (CFS)    (CFS)       ACRE-FT .0       66.      132.      197.     263.
  1.00        .7        .0       .057 O        .        .        .        .
     [BASIN DEPTH(FEET) =     .0]
  2.00       3.5        .1       .342 O        .        .        .        .
     [BASIN DEPTH(FEET) =     .0]
  3.00       8.1        .3       .995 O        .        .        .        .
     [BASIN DEPTH(FEET) =     .1]
  4.00      12.5        .6      1.991 OI       .        .        .        .
     [BASIN DEPTH(FEET) =     .1]
  5.00      16.1       1.0      3.258 OI       .        .        .        .
     [BASIN DEPTH(FEET) =     .2]
  6.00      19.2       1.4      4.748 O I      .        .        .        .
     [BASIN DEPTH(FEET) =     .3]
  7.00      22.3       1.9      6.454 O I      .        .        .        .
     [BASIN DEPTH(FEET) =     .4]
  8.00      31.3       2.6      8.855 O  I     .        .        .        .
     [BASIN DEPTH(FEET) =     .6]
  9.00      87.9       5.0     15.805 O        . I      .        .        .
     [BASIN DEPTH(FEET) =    1.1]
 10.00     198.3      18.7     31.216 . O      .        .        I        .
     [BASIN DEPTH(FEET) =    2.2]
 11.00     263.2      77.1     49.010 .        .O       .        .       I.
     [BASIN DEPTH(FEET) =    3.4]
 12.00     233.2     126.1     59.885 .        .      O.         .    I   .
     [BASIN DEPTH(FEET) =    4.2]
 13.00     170.5     141.3     62.924 .        .        .O   I   .        .
     [BASIN DEPTH(FEET) =    4.4]
 14.00     121.3     134.5     61.552 .        .      I O        .        .
     [BASIN DEPTH(FEET) =    4.3]
 15.00      88.3     118.6     58.391 .        . I    O .        .        .
     [BASIN DEPTH(FEET) =    4.1]
 16.00      66.9     102.4     54.787 .        I   O    .        .        .
     [BASIN DEPTH(FEET) =    3.8]
 17.00      52.5      87.1     51.296 .      I . O      .        .        .
     [BASIN DEPTH(FEET) =    3.6]
 18.00      42.6      73.5     48.181 .     I  O        .        .        .
     [BASIN DEPTH(FEET) =    3.3]
 19.00      35.3      61.8     45.508 .    I  O.        .        .        .
     [BASIN DEPTH(FEET) =    3.2]
 20.00      30.4      52.2     43.310 .  I  O .         .        .        .
     [BASIN DEPTH(FEET) =    3.0]
 21.00      27.3      46.9     41.470 .  I O  .         .        .        .
```

```
      [BASIN DEPTH(FEET) =    2.9]
 22.00     25.2     42.5       39.857 .  I O  .      .      .      .
      [BASIN DEPTH(FEET) =    2.8]
 23.00     23.9     38.7       38.478 . I O   .      .      .      .
      [BASIN DEPTH(FEET) =    2.7]
 24.00     23.0     35.5       37.314 . I O   .      .      .      .
      [BASIN DEPTH(FEET) =    2.6]
 25.00     20.3     32.4       36.188 . IO    .      .      .      .
      [BASIN DEPTH(FEET) =    2.5]
 26.00       .0     25.7       33.786 I  O    .      .      .      .
      [BASIN DEPTH(FEET) =    2.3]
 27.00       .0     20.5       31.876 I O     .      .      .      .
      [BASIN DEPTH(FEET) =    2.2]
 28.00       .0     16.3       30.356 IO      .      .      .      .
      [BASIN DEPTH(FEET) =    2.1]
 29.00       .0     13.0       29.147 IO      .      .      .      .
      [BASIN DEPTH(FEET) =    2.0]
 30.00       .0     11.6       28.131 IO      .      .      .      .
      [BASIN DEPTH(FEET) =    2.0]
 31.00       .0     11.1       27.190 IO      .      .      .      .
      [BASIN DEPTH(FEET) =   1.9]
 32.00       .0     10.6       26.290 IO      .      .      .      .
      [BASIN DEPTH(FEET) =   1.8]
 33.00       .0     10.2       25.429 IO      .      .      .      .
      [BASIN DEPTH(FEET) =   1.8]
 34.00       .0      9.7       24.606 IO      .      .      .      .
      [BASIN DEPTH(FEET) =   1.7]
 35.00       .0      9.3       23.818 IO      .      .      .      .
      [BASIN DEPTH(FEET) =   1.7]
 36.00       .0      8.9       23.065 IO      .      .      .      .
      [BASIN DEPTH(FEET) =   1.6]
 37.00       .0      8.5       22.345 IO      .      .      .      .
      [BASIN DEPTH(FEET) =   1.6]
 38.00       .0      8.2       21.656 O       .      .      .      .
      [BASIN DEPTH(FEET) =   1.5]
 39.00       .0      7.8       20.997 O       .      .      .      .
      [BASIN DEPTH(FEET) =   1.5]
 40.00       .0      7.5       20.367 O       .      .      .      .
      [BASIN DEPTH(FEET) =   1.4]
 41.00       .0      7.1       19.764 O       .      .      .      .
      [BASIN DEPTH(FEET) =   1.4]
 42.00       .0      6.8       19.188 O       .      .      .      .
      [BASIN DEPTH(FEET) =   1.3]
 43.00       .0      6.5       18.636 O       .      .      .      .
      [BASIN DEPTH(FEET) =   1.3]
 44.00       .0      6.2       18.109 O       .      .      .      .
      [BASIN DEPTH(FEET) =   1.3]
 45.00       .0      6.0       17.605 O       .      .      .      .
      [BASIN DEPTH(FEET) =   1.2]
 46.00       .0      5.7       17.122 O       .      .      .      .
      [BASIN DEPTH(FEET) =   1.2]
 47.00       .0      5.5       16.661 O       .      .      .      .
      [BASIN DEPTH(FEET) =   1.2]
 48.00       .0      5.2       16.219 O       .      .      .      .
      [BASIN DEPTH(FEET) =   1.1]
---------------------------------------------------------------------------
```

CHAPTER 5.
A SYNTHETIC UNIT HYDROGRAPH MODEL

Introduction

In this chapter, a synthetic unit hydrograph methodology is developed and the associated computer code presented for use on microcomputers. Such a synthetic runoff hydrograph approach may be used for generating estimates of the peak flow rates needed for the design of flood control channels, and estimates for a time distribution of runoff volume for the design of detention basins (Chapter 4).

Background

The unit hydrograph method is a synthetic hydrograph approach initially advanced by L. K. Sherman (1932). The keystone of the method is the assumption that watershed discharge is related to the total volume of runoff, and that the time factors which affect the unit hydrograph shape are invariant. The basic unit hydrograph theory was extended by F. F. Snyder (1938) to transpose storm rainfall-runoff relationships from gaged watersheds to hydrologically and geographically similar watersheds which lack rainfall-runoff stream gage data. The basic assumptions used in this later work are that the watershed rainfall-runoff relationships are functions of watershed area, slope and certain shape factors. The method is used to estimate a time distribution of runoff accumulating at the watershed downstream point of concentration when stream gage data is either unavailable or inadequate to provide a sound statistical analysis.

To determine the rainfall-runoff relationships to be transposed to ungaged watersheds, stream gage records are studied for various types and sizes of gaged watersheds. For example, the U.S. Army Corps of Engineers (Los Angeles Office) has determined several runoff time-distribution patterns for watersheds in the State of California. These relationships provide a basis for transposing to ungaged watersheds a characteristic time distribution of runoff which is the average distribution for several similar watershed basins. This approach is considered applicable when watersheds are physiographically and hydrologically similar. In Southern California, the counties of Orange, Riverside, and San Bernardino (which together represent a vast spectrum of flood control conditions) have successfully utilized this approach in the development of county-wide flood control facilities. Because the methodology used in the California watersheds would apply elsewhere (with appropriate modifications for local watershed rainfall-runoff relationships), the text and supplemental computer programs are based on the California rainfall-runoff relationships with the intent that the programs can be easily modified by the user to accomodate other regional watershed runoff tendencies.

Although there are several theoretical shortcomings associated with the unit hydrograph approach (such as the assumption of a linear system where runoff hydrographs resulting from a unit period of effective rainfall can be directly summed), the general approach continues to be widely used throughout the United States as a runoff synthesis method for ungaged watersheds. This chapter will examine the general unit hydrograph approach and several currently used variations of the method. A FORTRAN computer program is provided for a severe design storm condition (Hromadka et al., 1983a) which is readily adaptable to currently available microcomputers.

The unit hydrograph approach involves several assumptions which are imprecise approximations of the corresponding hydrologic processes. These basic assumptions are that (1) the critical storm rainfall pattern is uniformly distributed throughout the watershed, (2) there exists a direct proportionality between watershed runoff and the effective rainfall volume, (3) for any volume of effective rainfall occuring within a specified duration, the resulting runoff hydrograph is of a constant duration, and (4) the basin unit hydrograph is invariant throughout the critical design storm. The requirement that the watershed runoff is proportional to the effective rainfall has a direct analogy to a linear systems approach. Consequently, the unit hydrograph method can be considered a black-box modeling approach where the major characteristics of the model are determined by correlating the model output (runoff hydrograph) to the input data (rainfall records). Although the lumped-system model produces only approximations of the complex hydrologic characteristics of the watershed, its use continues to be widespread due to the ease of application. However, all of the several individual processes involved in the total watershed hydrologic system have an associated probabilistic relationship which represents the changing of the variable with respect to time and with respect to the fluctuation of the other variables. Thus, to incorporate the diverse probability effects into a lumped-system or a more sophisticated distributed parameter model, the general model has to be structured towards statistical interpretation of model results and an associated interval of confidence. For studies involving a severe critical design storm over a highly urbanized watershed, however, several of the variables may be eliminated from the hydrologic system model and the corresponding uncertainty is significantly reduced. For this type of condition, the unit hydrograph approach provides a runoff hydrograph which can be successfully used for flood control design purposes in watersheds which are hydrologically similar to gaged watersheds which have had unit hydrographs determined. A detailed analysis of the sensitivity of the unit hydrograph approach to the uncertainty associated to the precise definition of each model input parameter is given in Garen and Burges (1981). In that study, it was shown that the model output of the runoff hydrograph is very sensitive to variations in the input parameter set. The level of discretization of the watershed into smaller subareas to be linked together in a link-node model of the watershed is addressed in Zaghloul (1981) and includes the sensitivity effects of including the various routing processes in a complex model. The basic linearity assumption introduces theoretical difficulties which are discussed in Helweg et al. (1982). That study proposes methods for incorporating model memory and nonlinearity effects into a complex watershed model. Although the statistical component and the nonlinearity element should be included into the general runoff hydrograph generator, the difficulties associated with applying such sophistications in the general flood control study environment usually preclude their inclusion into a local agency's hydrology design criteria.

Terminology

In order to discuss the development of unit hydrographs for both gaged and ungaged watersheds, the following definitions are presented for the several terms:

Effective Rainfall: the total rainfall less infiltration losses, evaporation, transpiration, absorption, and detention. This part of the rainfall runs off the watershed surface in a relatively brief time period.

Unit Hydrograph: the unit hydrograph for a point of concentration on a watershed is a hydrograph showing the time distribution of runoff which results from a unit one inch of effective rainfall over the entire tributary watershed. The unit effective rainfall is assumed to occur as a constant rainfall with respect to both space and time throughout the unit duration. Figure 5.1 illustrates the general formulation of the unit hydrograph.

Distribution Graph: the distribution graph is a unit hydrograph whose ordinates are expressed in terms of percent of ultimate discharge. A distribution graph is generally developed as a block graph with each block representing its associated percent of unit runoff which occurs during the specified unit time period. The unit time used in the distribution graph is identical to the unit time specified for the unit hydrograph.

Summation Hydrograph: the summation hydrograph for a point of concentration is a hydrograph showing the time distribution of the rates of runoff that would result from a continuous series of unit effective rainfalls over the tributary watershed. The ordinates are expressed as rate of runoff in percent of the ultimate rate of runoff.

Lag: the watershed lag is the time (hours) from the beginning of a continuous series of unit effective rainfalls over the watershed to the instant when the rate of resulting runoff at the point of concentration reaches 50 percent of the ultimate rate of runoff. Another definition for lag is the time from the center of mass of the effective rainfall to the peak of the corresponding runoff hydrograph. Some hydrologists define lag as the time from the effective rainfall center of mass to the runoff hydrograph center of mass. In this chapter, however, the first definition of lag will be used for both the theoretical and computer program development.

Ultimate Discharge: the ultimate discharge (or the ultimate rate of runoff) is the maximum rate of watershed runoff which can result from a specified effective rainfall intensity. Ultimate discharge occurs when the rate of runoff on the summation hydrograph is equivalent to the rate of effective rainfall. For a unit effective rainfall intensity of one inch occuring in a unit interval of one hour, the ultimate discharge is 645 cfs for every square mile of watershed.

S-Graph: the S-graph is a summation hydrograph developed by plotting watershed discharge in percent of ultimate discharge as a function of time expressed in percent of watershed lag. Figure 5.1 illustrates each of the above unit hydrograph concepts and definitions.

Determination of Synthetic Distribution Graphs

In order to develop synthetic distribution graphs, adequate storm rainfall and watershed runoff information is required for several streams in the vicinity of the watershed under study. The distribution graphs for each of the gaged streams can be determined by trial and error attempts to duplicate the runoff hydrographs resulting from severe storm events. The derived distribution graphs are then verified by using them to reproduce runoff hydrographs for other major storm events. In this procedure, the watershed is assumed to be a linear system which ignores the variation of the unit hydrograph during the storm event. Models to approximately include memory and other nonlinearity effects have been studied by Helweg et al. (1982), among others.

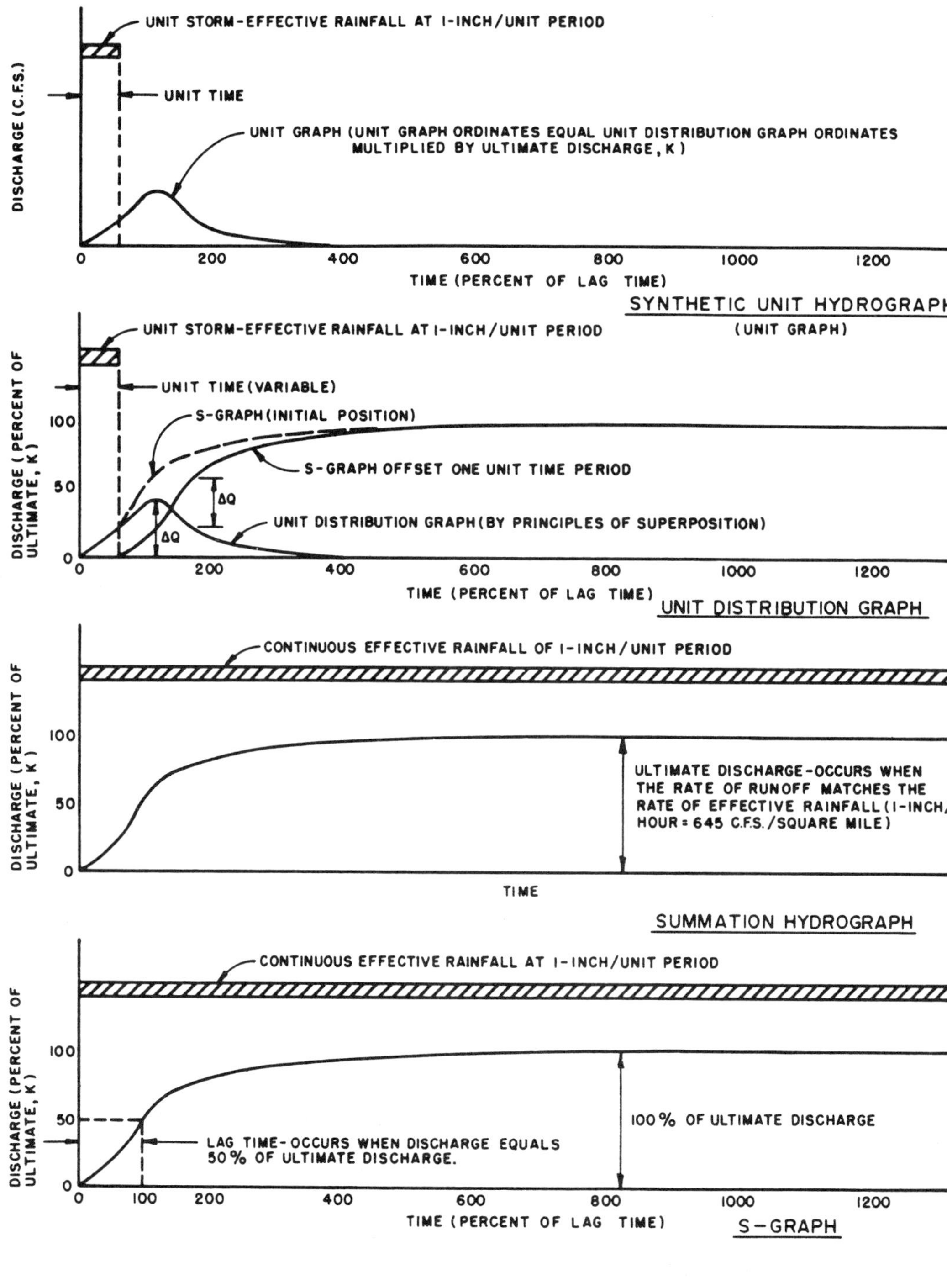

Fig. 5.1. Derivation of a synthetic unit hydrograph

It is assumed that the drainage areas within a given region are physiographically and hydrologically similar. Because no two drainage areas have identical hydrologic characteristics, the corresponding rainfall-runoff patterns are dissimilar and the distribution graphs will differ. Therefore, a direct transposition of distribution graphs between watersheds is usually not feasable. However, most distribution graphs exhibit certain characteristics which appear to be related to the empirical factor of watershed lag. Details for determining watershed lag for watersheds where the time distribution of runoff is known and for the use in developing synthetic distribution graphs are discussed in the following steps:

(1). Summation Hydrograph. The first step in determining watershed lag is the development of a characteristic summation hydrograph, which is the runoff hydrograph that results from a continuous and uniform sequence of unit effective rainfalls. The ordinates of the summation hydrograph are expressed as discharge in percent of ultimate discharge. The summation hydrograph for a point of concentration is also determined by adding a continuous series of identical distribution graphs each offset by one unit period. By examination, the time required to reach the ultimate discharge is equal to the length of the distribution graph base less one unit period.

(2). Lag. Watershed lag can be defined as the time from the beginning of unit effective rainfall to the instant that the summation hydrograph reaches 50 percent of ultimate discharge. When the lags determined from summation hydrographs for several gaged watersheds are correlated to the hydrologic characteristics of the watersheds, an empirical relationship is usually apparent. This relationship can then be used to estimate lag for ungaged watersheds. Analysis of watershed lags estimated for several types of watersheds indicate that a lag factor is expressed by

$$\text{lag} = C_t L^m L_c{}^m S^{-m/2} \qquad (5.1)$$

where

lag = watershed lag in hours
C_t = a constant
L = length of longest watercourse (miles)
L_c = length along longest watercourse, measured from the point of concentration to a point opposite the watershed area centroid
S = mean watershed slope along watercourse (feet/mile)
m = a constant

For Southern California watersheds, it is assumed that

$$\begin{aligned} n^* &= \text{a basin factor(see Fig. 5.2)} \\ m &= 0.38 \\ C_t &= 24n^* \end{aligned} \qquad (5.2)$$

The basin factors of Fig. 5.2 were determined by the U.S. Army Corps of Engineers by duplicating major storm runoff hydrographs from several Southern California watersheds. The relationship between watershed lag and equation (5.1) is illustrated in the plot of Appendix A. From (5.1), the synthetic unit hydrograph approach is a strong function of the watershed lag, and a wide range of results are generated by a variation of the basin factor.
(3). S-Graph. After lag factors are determined for several gaged watersheds, the next step in determining synthetic distribution graphs is the development of S-graphs, which are summation hydrographs modified so that the percent of ultimate discharge is plotted against time expressed in percent of watershed lag. The derivation of a S-graph is identical to the development of a summation hydrograph except that the relationships are now expressed with respect to watershed lag. Four S-graphs assumed applicable in Southern California watersheds are shown in Figs. 5.3a,b,c,d. These S-graphs are for use in watersheds classified as valley, foothill, mountain, or desert. According to the definitions, each S-graph reaches 50 percent of ultimate discharge at 100 percent of watershed lag.

(4). Application of Lag and S-Graphs. To use the S-graphs, a watershed lag is estimated from Figs. 5.1 and 5.2. A unit period is selected (usually at 15 to 25 percent of the watershed lag time) and amassed unit periods are expressed as accumulated percentages of the watershed lag. These percentages of lag are used for superimposing a block graph on the selected S-graph and the resulting block graph pattern is used in determining the accumulated mean percentage of ultimate discharge for each accumulated unit period. Finally, the incremental mean percentage of ultimate discharge for each unit period is estimated by a series of successive subtractions. The following section describes in detail the discussed procedure as applied in developing a watershed unit hydrograph.

Development of a Synthetic Unit Hydrograph

For watersheds where stream gage data is inadequate for statistical analysis, a synthetic unit hydrograph may be used to approximately describe the time distribution of runoff gathering at the point of concentration. A method for developing a synthetic unit hydrograph is described in the following steps:

n^* = 0.015

1. Drainage area has fairly uniform, gentle slopes
2. Most watercourses either improved or along paved streets
3. Groundcover consists of some grasses - large % of area impervious
4. Main water course improved channel or conduit

n^*= 0.020

1. Drainage area has some graded and non-uniform, gentle slopes
2. Over half of the area watercourses are improved or paved streets
3. Groundcover consists of equal amount of grasses and impervious area
4. Main watercourse is partly improved channel or conduit and partly greenbelt (see n = 0.025)

n^* = 0.025

1. Drainage area is generally rolling with gentle side slopes
2. Some drainage improvements in the area - streets and canals
3. Groundcover consists mostly of scattered brush and grass and small % impervious
4. Main watercourse is straight channels which are turfed or with stony beds and weeds on earth bank (greenbelt type)

n^* = 0.030 (Foothill Area)

1. Drainage area is generally rolling with rounded ridges and moderate side slopes
2. No drainage improvements exist in the area
3. Groundcover includes scattered brush and grasses
4. Watercourses meander in fairly straight, unimproved channels with some boulders and lodged debris

n^* = 0.040 (Foothill Area)

1. Drainage area is composed of steep upper canyons with moderate slopes in lower canyons
2. No drainage improvements exist in the area
3. Groundcover is mixed brush and trees with grasses in lower canyons
4. Watercourses have moderate bends and are moderately impeded by boulders and debris with meandering courses

n^* = 0.050 (Mountain Areas)

1. Drainage area is quite rugged with sharp ridges and steep canyons
2. No drainage improvements exist in the area
3. Groundcover, excluding small areas of rock outcrops, includes many trees and considerable underbrush
4. Watercourses meander around sharp bends, over large boulders and considerable debris obstruction

n^* = 0.200

1. Drainage area has comparatively uniform slopes
2. No drainage improvements exist in the area
3. Groundcover consists of cultivated crops or substantial growths of grass and fairly dense small shrubs, cacti, or similar vegetation
4. Surface characteristics are such that channelization does not occur

Fig. 5.2. Basin n parameter descriptions*

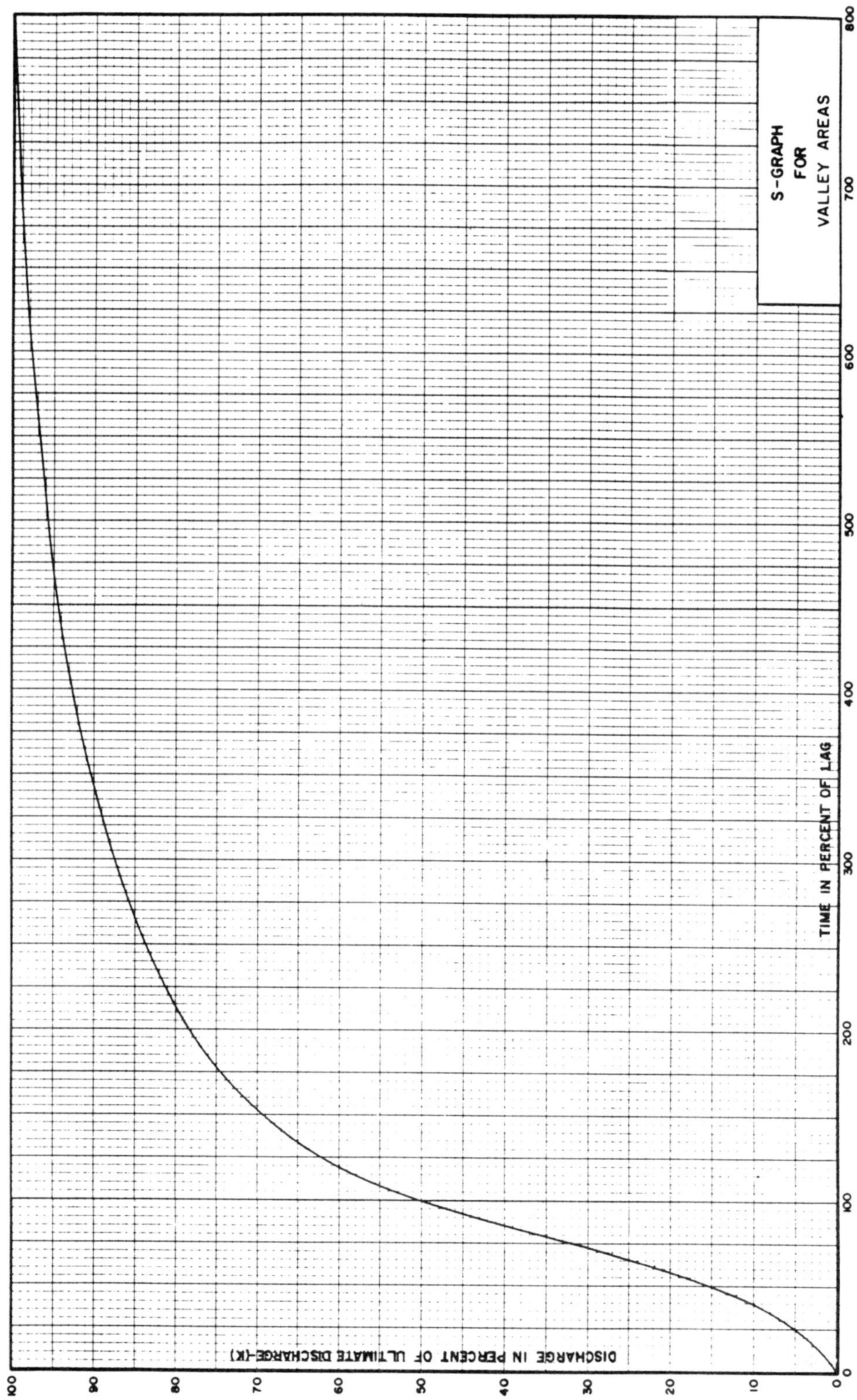
S-GRAPH
FOR
VALLEY AREAS
TIME IN PERCENT OF LAG
0
100
200
300
400
500
600
700
800
DISCHARGE IN PERCENT OF ULTIMATE DISCHARGE-(K)
0
10
20
30
40
50
60
70
80
90
100

Fig. 5.3a.

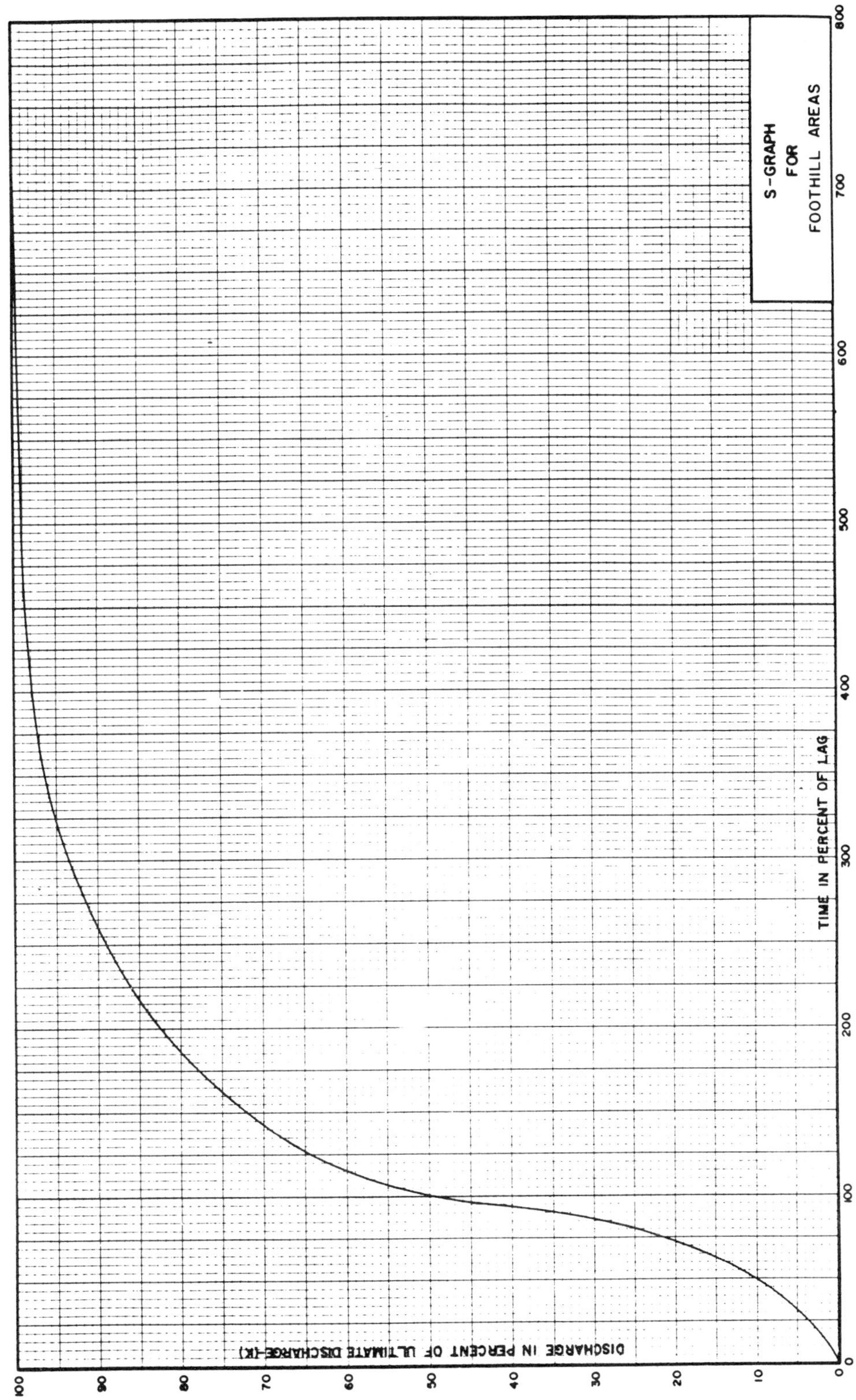
S-GRAPH
FOR
FOOTHILL AREAS
TIME IN PERCENT OF LAG
0 100 200 300 400 500 600 700 800
DISCHARGE IN PERCENT OF ULTIMATE DISCHARGE-(K)
0 10 20 30 40 50 60 70 80 90 100

Fig. 5.3b.

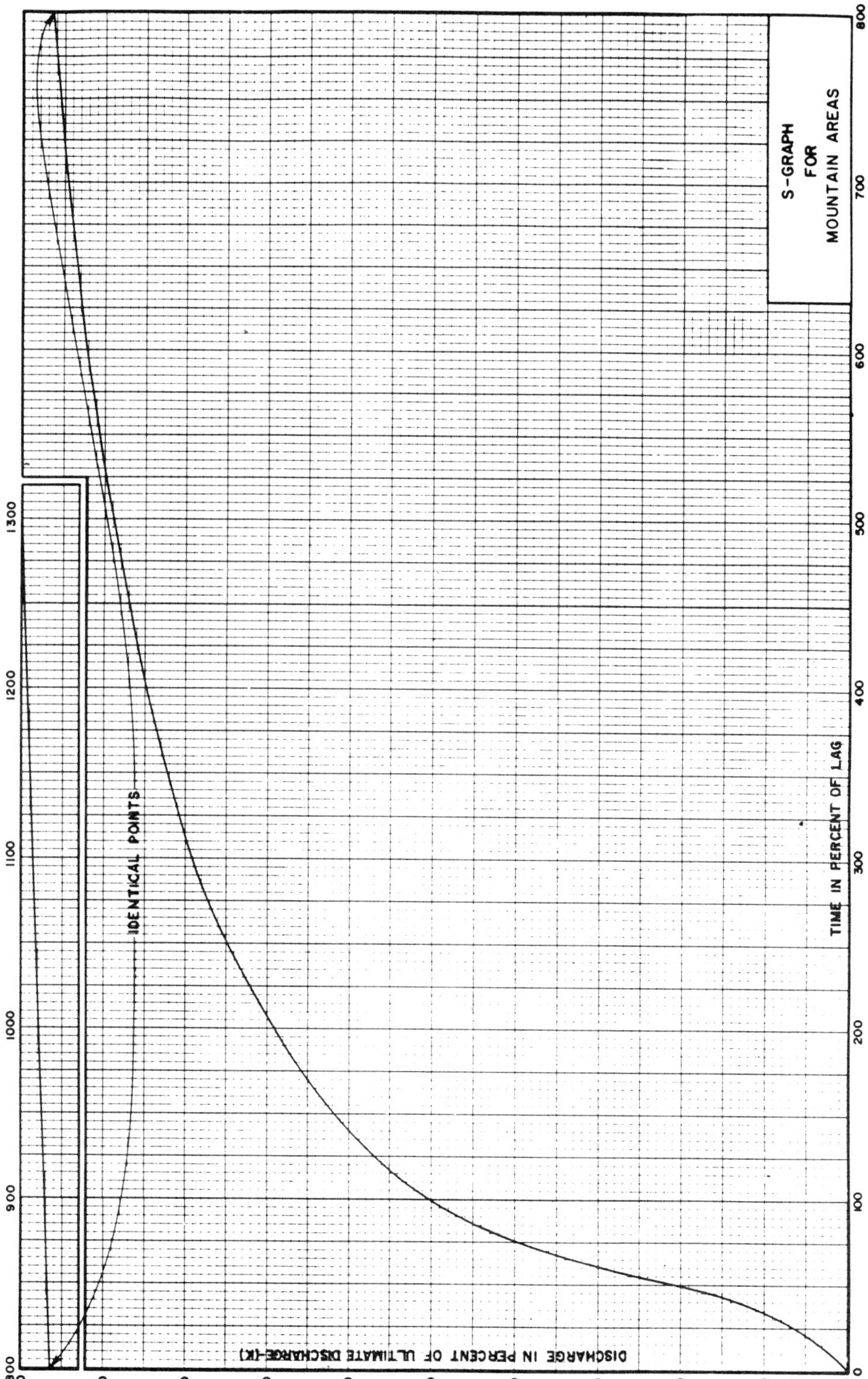
S-GRAPH
FOR
MOUNTAIN AREAS
TIME IN PERCENT OF LAG
DISCHARGE IN PERCENT OF ULTIMATE DISCHARGE-(K)
IDENTICAL POINTS
0 100 200 300 400 500 600 700 800
800 900 1000 1100 1200 1300
0 10 20 30 40 50 60 70 80 90 100

Fig. 5.3c.

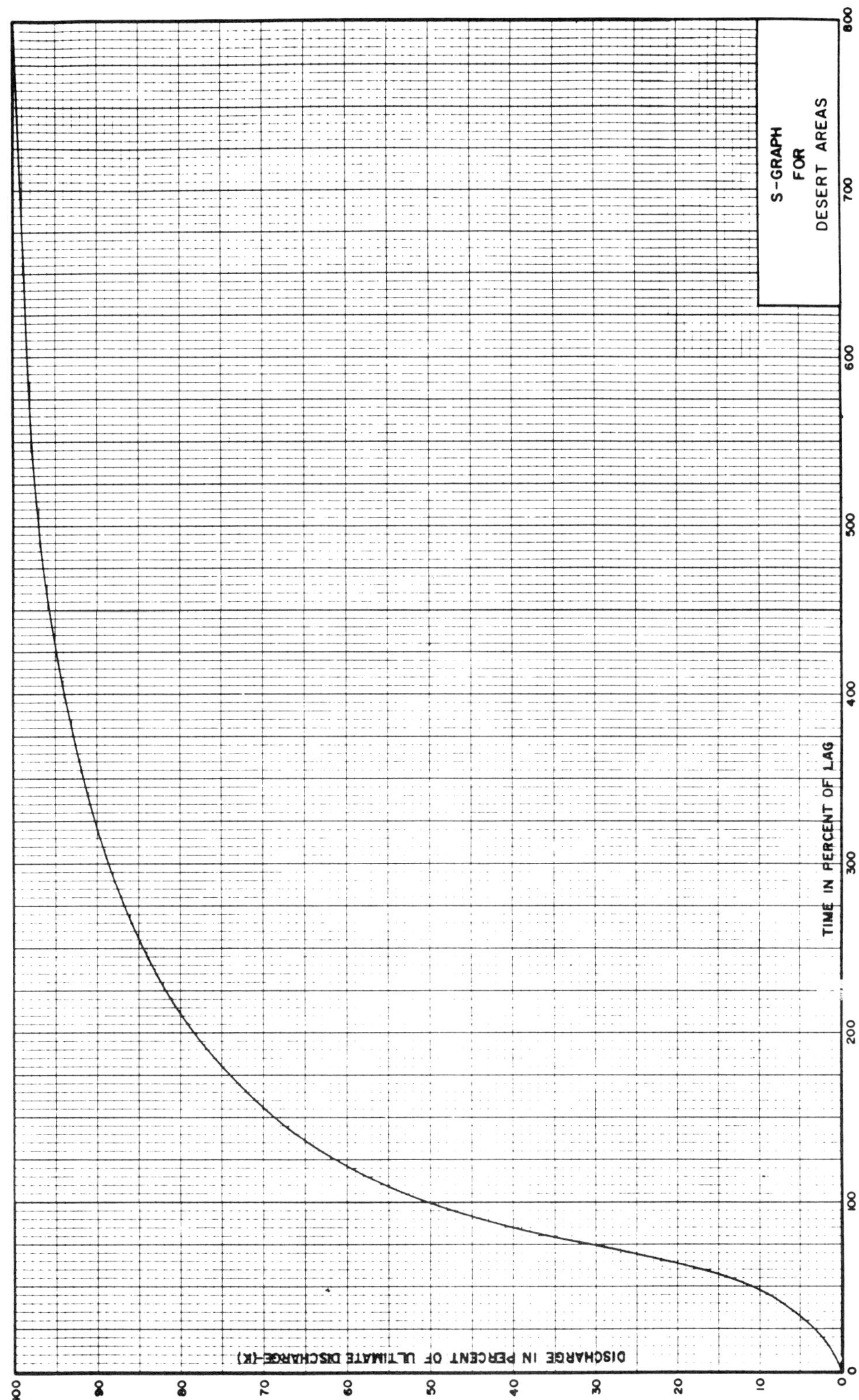
S-GRAPH
FOR
DESERT AREAS
TIME IN PERCENT OF LAG
DISCHARGE IN PERCENT OF ULTIMATE DISCHARGE-(K)
0
100
200
300
400
500
600
700
800
0
10
20
30
40
50
60
70
80
90
100

Fig. 5.3d.

(1). Estimate the watershed lag using topographic information, equation (5.1), and the appropriate basin factors (such as Fig. 5.2 for Southern California watersheds).

(2). Select a unit period duration. This unit period will be the time base for describing unit rainfalls and the corresponding runoff hydrographs. Usually, 15 to 25 percent of the watershed lag is adequate for computational purposes.

(3). Select a watershed S-graph which is determined from a set of gaged watersheds which are physiographically and hydrologically similar to the study watershed.

(4). The S-graph can be approximated by a block graph where the base of each block is the selected unit period (expressed in percentage of watershed lag time) and the ordinate of each block is the time-averaged percentage of ultimate discharge (from the S-graph) for the corresponding unit period. The area of each block equals the area under the S-graph for each respective unit period.

(5). The unit distribution graph is approximated by calculating the difference between the ordinates of the block graph used to approximate the watershed S-graph. This procedure is analogous to calculating the difference between the ordinates of two S-graphs which have been offset by one unit period.

(6). The final step in developing the synthetic unit hydrograph is to multiply the ordinates of the distribution block graph by the ultimate discharge defined by

$$K = 645A/T \tag{5.3}$$

where

K = the watershed ultimate discharge (cfs)
A = drainage area (square miles)
T = unit time period (hours)

A widely used unit hydrograph is the S.C.S. dimensionless unit hydrograph of Fig. 5.4. From the figure, the concept of time to peak (T_p) is used and is shown to be geometrically related to the watershed lag time. The County of San Diego, California (1975) uses the empirical watershed lag formulas of (5.1) and (5.2) plus the relationship

$$T_p = (0.862)\text{lag} \tag{5.4}$$

Equation (5.4) is based on the geometric relationships for the hydrographs shown in Fig. 5.4 which includes the triangular version of the S.C.S. curvilinear unit hydrograph. From Fig. 5.4, it is assumed that

$$\begin{aligned} T_r &= (1.67)T_p \\ T_b &= (2.67)T_p \end{aligned} \tag{5.5}$$

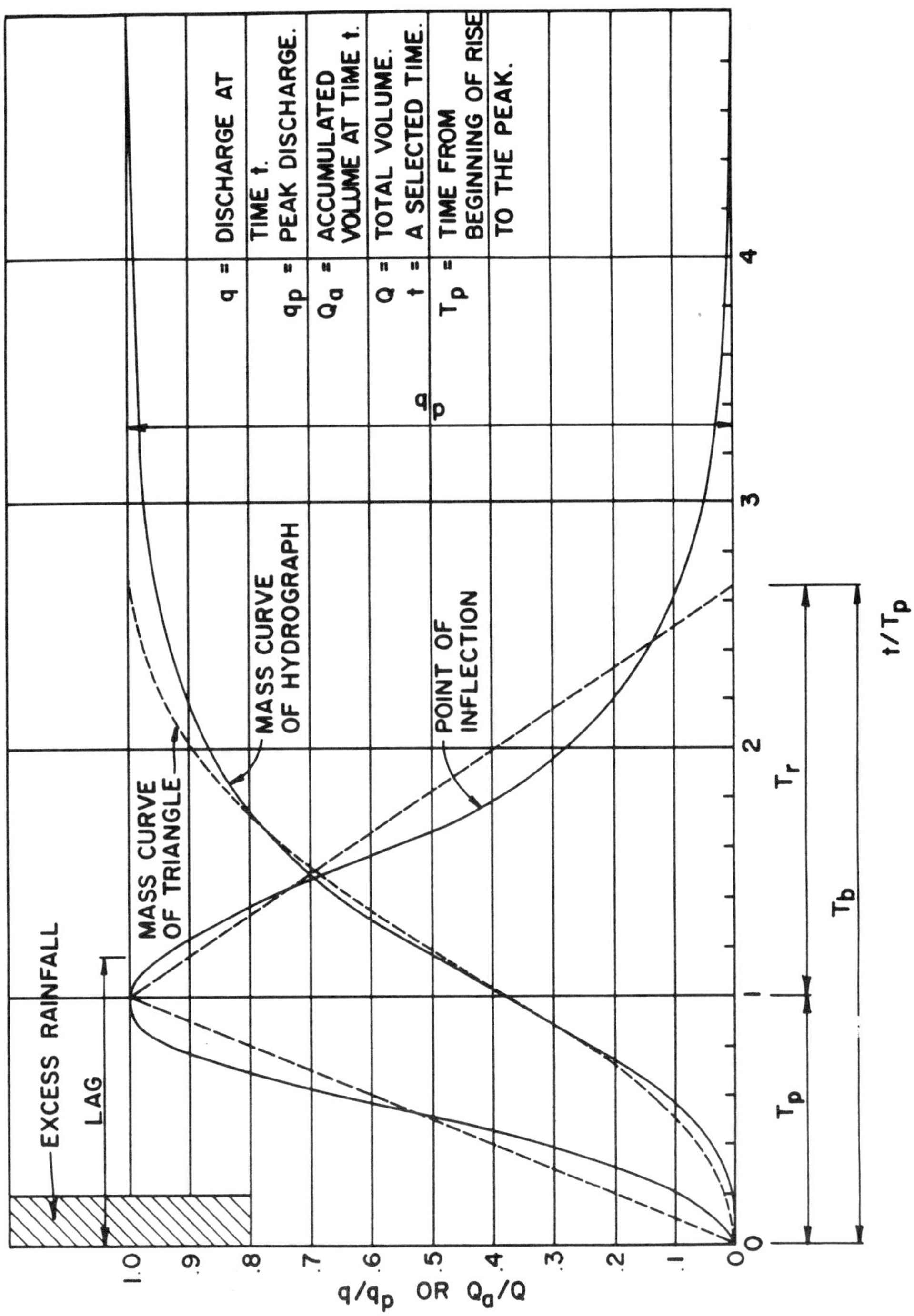

Fig. 5.4. S.C.S. dimensionless unit hydrograph and equivalent triangular hydrograph

A tabulation of the S.C.S. dimensionless unit hydrograph ratios is given in Table 5.1. From the table, it is seen that whatever the condition or classification of the watershed, the ratio of the time to peak to total unit hydrograph duration is a constant. Additionally, it is seen that the volume of runoff under the rising limb of the unit hydrograph is a constant 37.5 percent of the total runoff volume, and that the peak rate factor associated to the unit rainfall is a constant value of 484. The variability of the S.C.S. unit hydrograph with respect to other peak rate factors is investigated by McCuen and Bondelid (1983). In their study, methods to estimate a peak rate factor for ungaged watersheds is proposed using a gamma function distribution for the unit hydrograph and the estimation of approximate watershed storage effects in the determination of the volume of runoff defined under the rising limb. As with many other unit hydrograph schemes, the assumption of linearity is once again utilized to develop a runoff hydrograph. Watershed peak rate factors are determined by time-area analysis and attempts to duplicate runoff hydrographs. The study concluded that the peak rate factor of 484 may be inappropriate for many watersheds.

Possibly, more insight into the mechanics of the unit hydrograph is gained by working with the Espey and Altman (1978) approach. This version of the unit hydrograph involves a set of five parameters which are defined according to Fig. 5.5. The variable relationships are of the form

$$T_p = (3.1)L^{0.23}S^{-0.25}I^{-0.18}K^{1.57}$$

$$T_B = (125890)AQ^{-0.95}$$

$$Q = (31620)A^{0.96}T_p^{-1.07}$$

$$W_{50} = (16220)A^{0.93}Q^{-0.92}$$

$$W_{75} = (3240)A^{0.79}Q^{-0.78} \tag{5.6}$$

where

L = main channel length (feet)
H = drop in elevation along channel
S = channel slope
I = impervious area fraction
A = watershed area (square miles)
T_p = time to peak (minutes)
T_B = base time (minutes)
K = a dimensionless channel conveyance factor
W_{50}, W_{75} = width at 50 percent and 75 percent Q

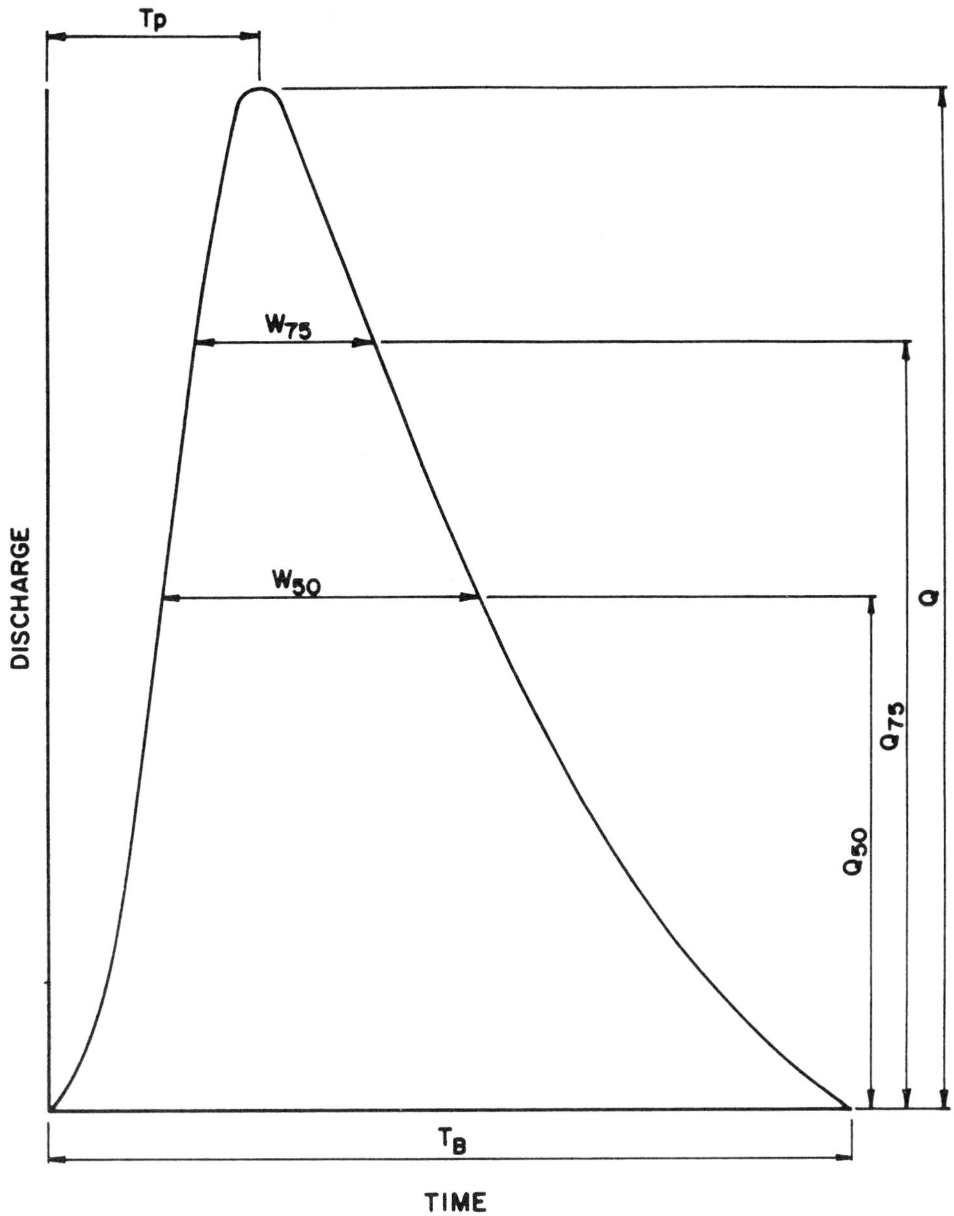

Fig. 5.5. *Definition of Espey 10-minute unit hydrograph parameters*

The Espey unit hydrograph is sensitive to the channel conveyance factor (K) in a fashion analogous to the sensitivity of the general unit hydrograph to the basin n^* factor. Both of these methods require some sort of tuning in order to evaluate their respective parameters.

TABLE 5.1. RATIOS FOR THE S.C.S. UNIT HYDROGRAPH

Time Ratio	Discharge Ratio	Mass Curve Ratio
0.00	0.000	0.000
0.10	0.030	0.001
0.20	0.100	0.006
0.30	0.190	0.012
0.40	0.310	0.035
0.50	0.470	0.065
0.60	0.660	0.107
0.70	0.820	0.163
0.80	0.930	0.228
0.90	0.990	0.300
1.00	1.000	0.375
1.10	0.990	0.450
1.20	0.930	0.522
1.30	0.860	0.589
1.40	0.780	0.650
1.50	0.680	0.700
1.60	0.560	0.751
1.70	0.460	0.790
1.80	0.390	0.822
2.00	0.280	0.871
2.20	0.207	0.908
2.40	0.147	0.934
2.60	0.107	0.953
2.80	0.077	0.967
3.00	0.055	0.977
3.20	0.040	0.984
3.40	0.029	0.989
3.60	0.021	0.993
3.80	0.015	0.995
4.00	0.011	0.997
4.50	0.005	0.999
5.00	0.000	1.000

Intensity-Duration Curves for Design Storm Point Precipitation

Rainfall intensity versus storm duration data is required for use in developing a synthetic design storm. Intensity-duration curves can be developed for a watershed by estimating the area-averaged one hour point precipitation values from synthesized local rain gage data, or from reliable point precipitation maps. Using Fig. 5.6, the one hour point precipitation value is plotted,

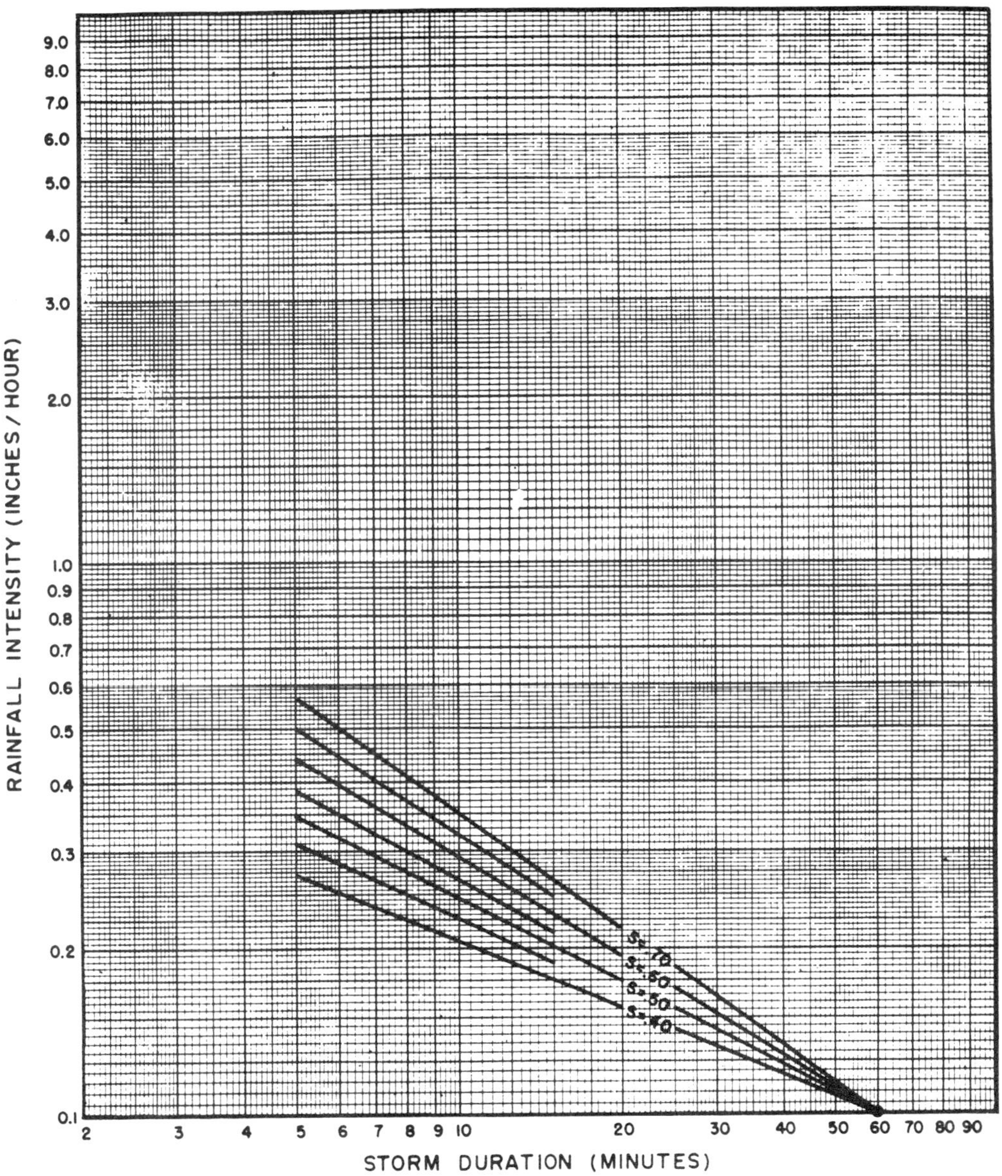

DESIGN STORM FREQUENCY = _____ YEARS

ONE HOUR POINT RAINFALL = _____ INCHES

LOG-LOG SLOPE = _____

PROJECT LOCATION = __________________

Fig. 5.6. Intensity-duration curve calculation sheet

and a straight line is drawn with the appropriate slope. Generally, slopes vary from about 0.30 to 0.80.

Area-Averaged Point Rainfall

Due to the potential variation of precipitation depth (for a given duration) throughout a watershed, approximate methods are required in order to estimate an area-averaged point rainfall for the entire watershed. The three most often used methods are the arithmetic method, the Thiessen polygon method, and the isohyetal method. The application of these methods is shown in Fig. 5.7.

The arithmetic method is a simple average of point rainfall values. It is the least accurate of the three approaches for area-averaging.

The Thiessen polygon method utilizes an area-averaged value based on polygons associated to each rain gage in the vicinity of the watershed. The polygons are determined by constructing perpendicular bisectors to lines connecting the several rain gages.

Finally, the isohyetal method is a more precise approach in which a map of the interpolated point rainfall isohyetals are plotted on the watershed and the area-averaged point rainfall is directly estimated by contour integration.

Synthetic Critical Storm Patterns

The use of computer watershed simulation techniques for hydrologic analysis has become widespread with the advent of inexpensive digital computer availability. Generally speaking, the basis of the hydrologic modeling approach is to study the effects of a severe storm event (or design storm) upon the watershed. Consequently, the decision as to what critical storm to use for study purposes has a significant impact upon the ultimate design objective in providing flood control.

The critical storm approach falls into two categories: (1) a severe storm pattern of record, and (2) a synthetic critical storm pattern. The critical storm pattern of record is a historical rainfall event which is assumed to be associated with the severe flooding event. The storm may have occurred within the vicinity of the watershed, or may have occurred elsewhere but is assumed to be transposable to the study watershed without violating the assumptions of uniformity of hydrologic or geographic characteristics. For example, the U.S. Army Corps of Engineers (Los Angeles, Calif.) utilize for study purposes a critical storm pattern based on a 1943 thunderstorm (Fig. 5.8). From Fig. 5.8, the 3-hour duration storm is composed of twelve 15-minute unit durations of precipitation depths. These unit rainfall values are estimated from peak 15-, 30-, 60-, and 180- minute duration

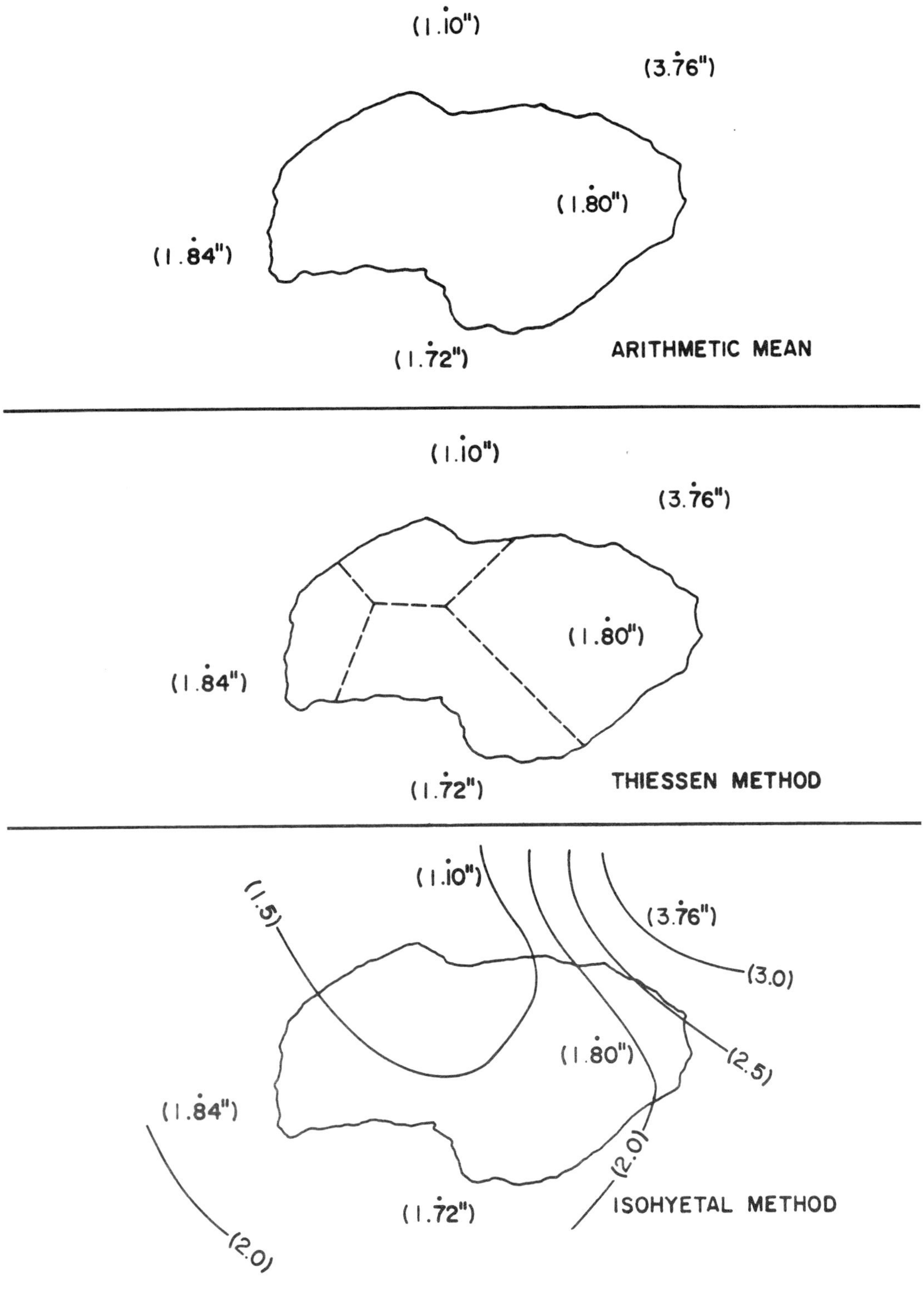

Fig. 5.7. Point rainfall area-averaging methods

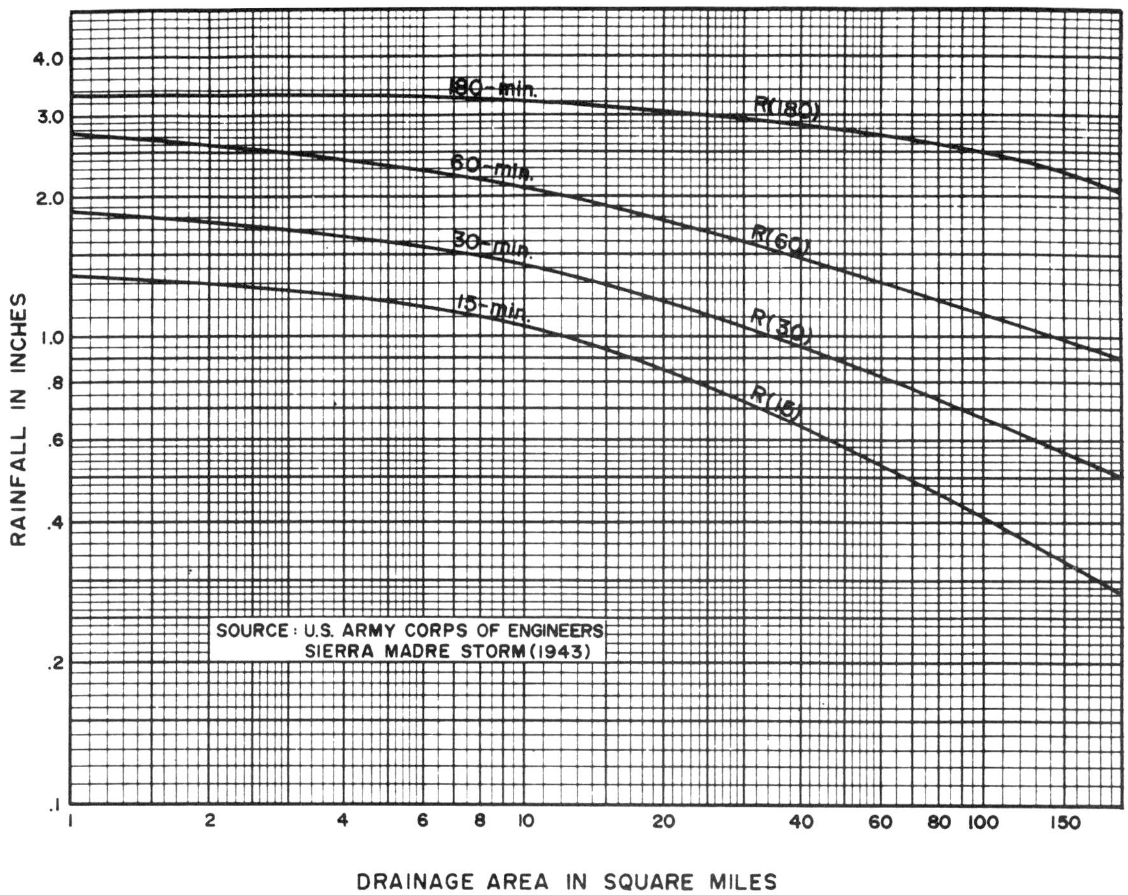

HYETOGRAPH COMPUTATION

UNIT PERIOD	AMOUNT
1	.07 (R(180) - R(60))
2	.05 (R(180) - R(60))
3	.11 (R(180) - R(60))
4	.05 (R(180) - R(60))
5	.20 (R(180) - R(60))
6	.22 (R(180) - R(60))
7	.14 (R(180) - R(60))
8	.16 (R(180) - R(60))
9	.48 (R(60) - R(30))
10	.52 (R(60) - R(30))
11	1.00 (R(15))
12	1.00 (R(30) - R(15))

UNIT PERIOD = 15 MINUTES

Fig. 5.8. Critical storm depth-area-duration curves

precipitation-depth values as a function of watershed area. The storm pattern is based upon a 3.3-inch total storm rainfall. In transposing the storm to another watershed, the local 3- hour duration precipitation-depth is estimated (for the desired return frequency) and the adjusted storm pattern is obtained by a simple ratio. For a depth of 3.3 inches, the resulting storm pattern is shown in Fig 5.9. A difficulty with this storm pattern is that the precipitation-depth ratio of interior durations to the total 3-hour duration may not be appropriate for the local region of interest, and an unacceptably high (or low) precipitation depth may be assigned to the interior unit rainfalls. For example, the design objective may be to provide flood protection for a 100-year return frequency rainfall for the critical duration of every hydraulic structure within a watershed. Then using a 100- year 3-hour depth with the given storm pattern may possibly result in a 150- year (or 50- year) return frequency rainfall depth to be assigned for some critical duration. Designing the structure for such a runoff rate would result in an overdesign (or underdesign).

The second category of critical storm patterns is the synthetic critical storm pattern. This approach is an attempt to accomodate some of the concerns which are associated with using critical storm patterns of record. Several synthetic storm patterns have been proposed, such as Keifer and Chu (1957), Huff (1967), Terstriep and Stall (1974), Pilgrim and Cordery (1975), among others. For example, Hershfield (1962) developed a normalized time distribution of total mass rainfall versus storm duration. This generalized distribution is given in Table 5.2.

TABLE 5.2. HERSHFIELD MASS RAINFALL DISTRIBUTION

Storm Duration	Mass Rainfall
0.00	0.00
0.10	0.06
0.20	0.12
0.30	0.20
0.40	0.29
0.45	0.34
0.50	0.45
0.55	0.63
0.60	0.73
0.65	0.81
0.70	0.86
0.80	0.94
0.90	0.99
1.00	1.00

A widely used set of critical storm patterns was developed by the U.S. Department of Agriculture, Soil Conservation Service (S.C.S., 1973). The synthetic storm patterns are presented in normalized mass rainfall fashion for both a 6- hour and 24- hour

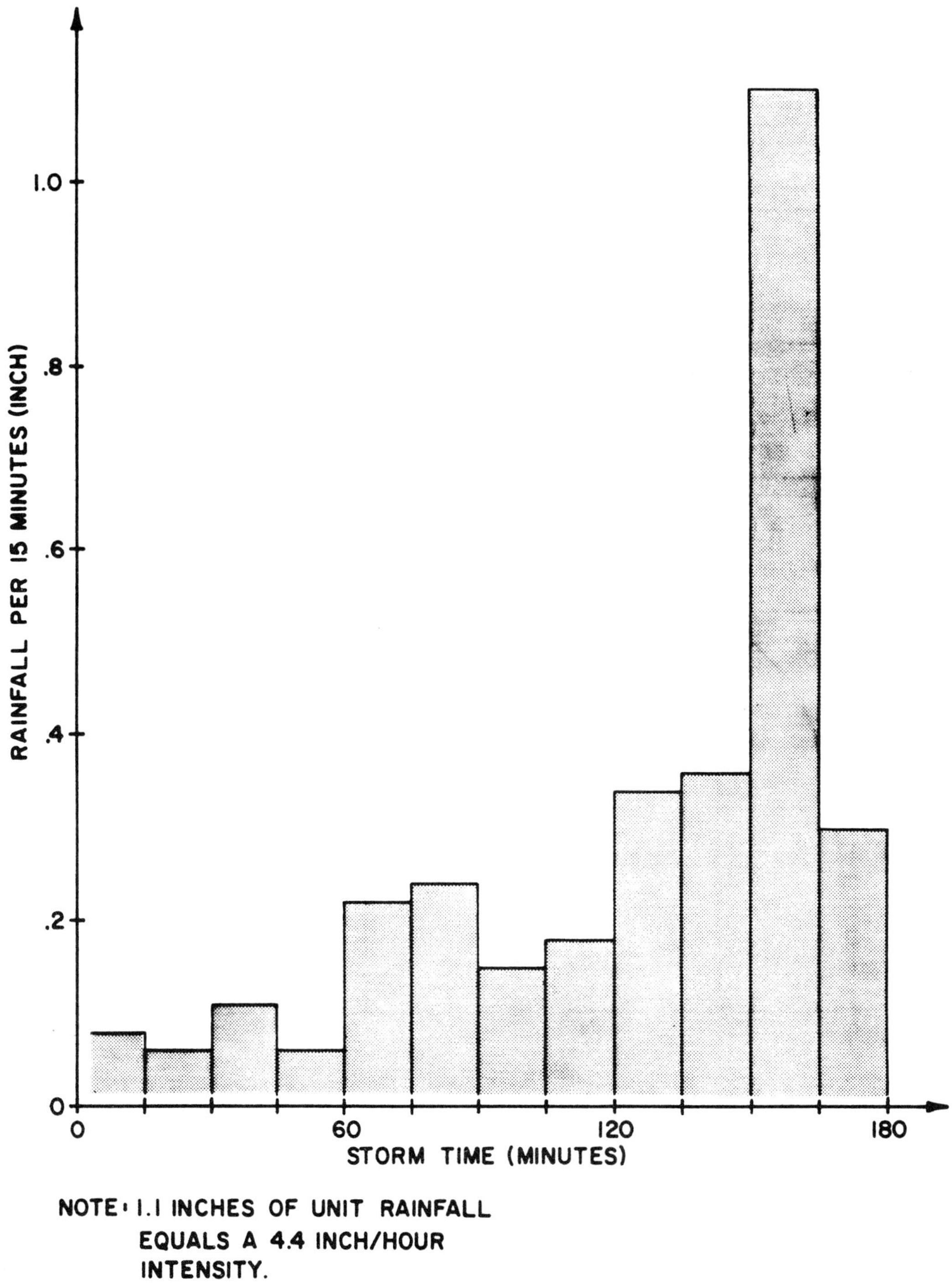

Fig. 5.9. 3.3-inch depth critical storm pattern (10 square mile area)

duration critical storm event. The 24- hour storm pattern is further classified as a type I or type II, in order to better represent different storm precipitation intensity characteristics as averaged over broad regions of the United States. The S.C.S. developed the dimensionless synthetic storm patterns using the U.S. Weather Service rainfall frequency reports. The patterns are based upon regionally averaged rainfall gage data for watershed areas less than approximately 400 square miles, with durations to 24 hours, and return frequencies from 1 year to 100 years.

These critical storm patterns represent an average of generalized precipitation depth-duration-frequency relationships and, consequently, may be inappropriate for many hydrologic studies where the objective is to design flood control facilities to accomodate the maximum precipitation depth (with a specified return frequency) for the critical duration of each hydraulic structure within the system.

TABLE 5.3. S.C.S. SYNTHETIC STORM PATTERN
(TYPE I, 24 HOUR STORM)

Storm Hour	Mass Rainfall
0.0	0.000
2.0	0.035
4.0	0.076
6.0	0.125
6.0	0.156
8.0	0.194
8.5	0.219
9.0	0.254
9.5	0.303
9.75	0.362
10.0	0.515
11.0	0.624
12.0	0.682
13.0	0.727
14.0	0.767
16.0	0.830
20.0	0.926
24.0	1.000

The storm patterns are constructed by subtracting the peak 30-minute depth from the peak 1- hour depth, and subtracting the peak 1- hour depth from the peak 1.5- hour depth, and so forth until a set of 30-minute unit rainfalls results. The unit rainfalls are arranged such that a nesting of the peak unit rainfalls are formed where the peak 30 minutes occurs within the peak 1- hour, the peak hour occurs within the peak 1.5 hours, and so forth until the 24- hour storm pattern is defined. Because all the critical precipitation depths are contained within the storm pattern, the synthetic storm patterns are generally assumed appropriate for design study purposes for both small and large watersheds. The dimensionless S. C. S. 24- hour type I pattern is presented in Table 5.3.

An examination of both the Hershfield and the S.C.S. synthetic storm patterns indicate that neither pattern guarantees a critical precipitation depth for an arbitrary critical duration. That is, both patterns define each interior storm duration as a fixed percentage of total storm mass rainfall. The Composite Method for developing a synthetic storm pattern is analogous to the approach used by the S.C.S. except that the local rain gage data is used to determine the unit rainfalls. This method simply uses the local rainfall intensity-duration curves with the desired return frequency, and incremental unit rainfalls are determined by successive subtractions. The unit rainfalls are then arranged in some nested pattern with the peak intensities usually defined to occur at 0.50, 0.67, or some percentage of the total storm duration. Figures 5.10 a,b,c show a typical composite storm pattern which is used for design hydrology purposes in the County of San Bernardino, California (Hromadka and Guymon, 1983b).

Depth-Area Relationships

A critical storm pattern can be directly applied towards estimating peak flow rates and runoff volumes associated to small watersheds where stream gage records are inadequate for statistical frequency analysis. As the watershed area increases in size, however, studies indicate (Miller et al., 1973, Myers and Zehr, 1980) that the area-averaged point rainfall values used in the design storm pattern should be adjusted. For example, the NOAA Atlas 2 includes point rainfall depth-area reduction relationships (Fig. 5.11) for 30- minute, 1-, 3-, 6-, and 24- hour point rainfall values as a function of watershed size.

Modified Composite Storm Pattern

The composite storm pattern approach can be modified to include the effects of depth-area adjustment. However, due to the adjustment factors being available for only a few durations, an interpolation procedure is required for the remaining durations within the storm pattern. One approach is to plot the adjusted point rainfall values on standard log-log paper and assume a straight line relationship (see Fig. 5.12). Other relationships can be used, but the log-log plot is convenient in that the resulting mass rainfall plot is simple to approximate numerically. Unit rainfalls are determined for the modified composite storm pattern by successive subtractions of unit durations along the adjusted mass rainfall plot. The unit rainfalls are then arranged as is shown in Fig. 5.13.

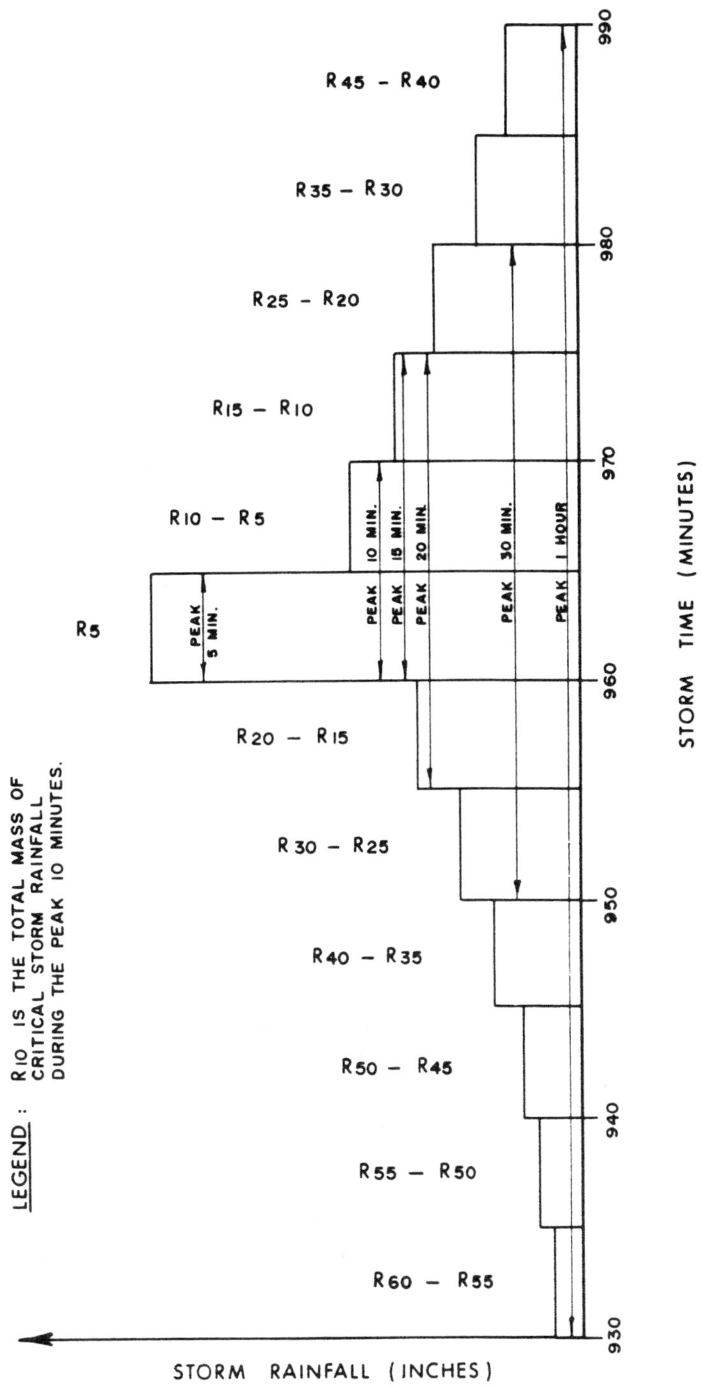

Fig. 5.10a. *Synthetic critical storm peak 1-hour pattern*

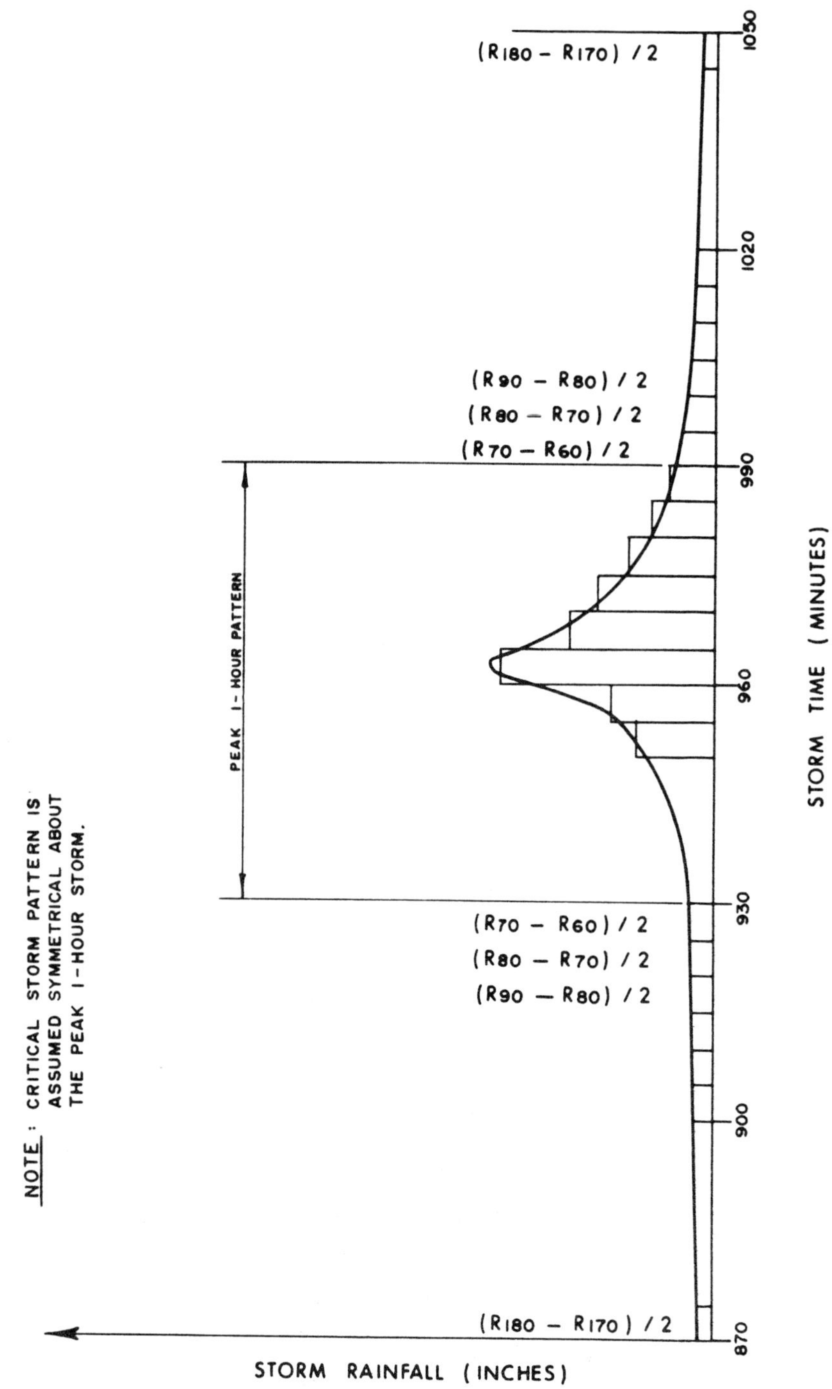

Fig. 5.10b. Synthetic critical storm peak 3-hour pattern

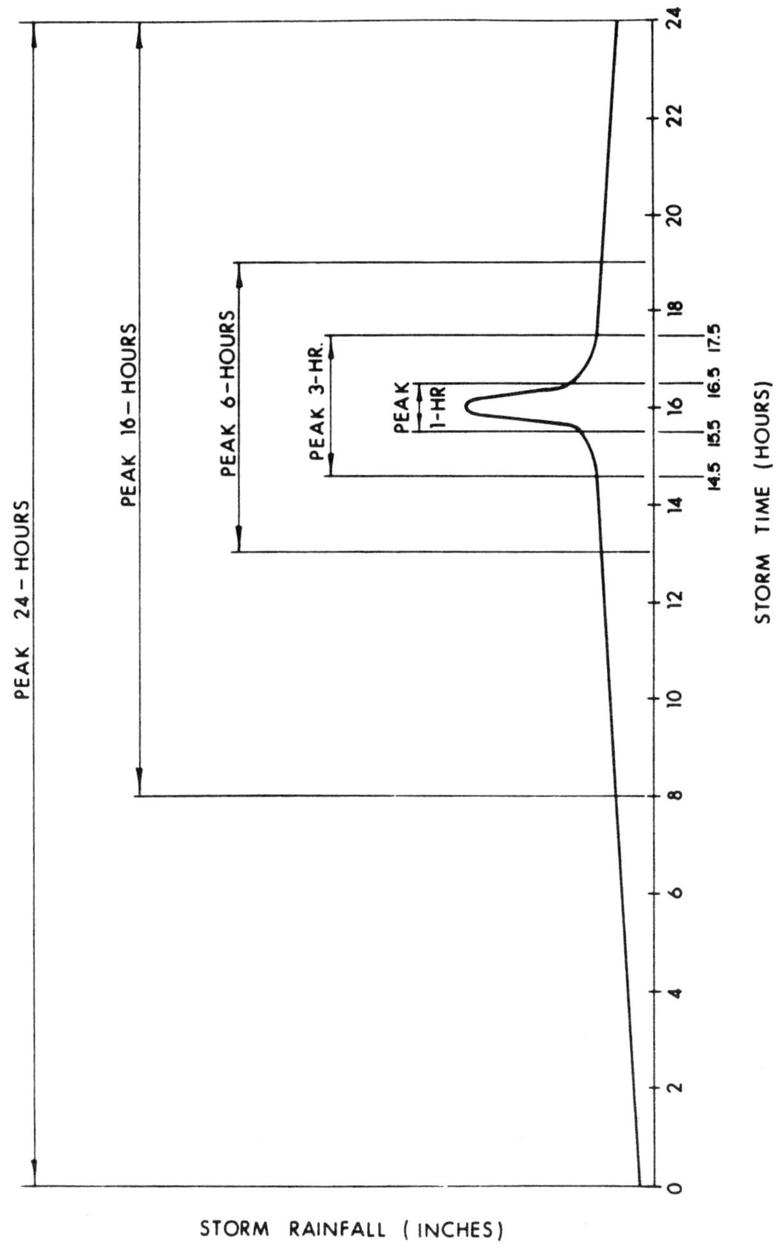

Fig. 5.10c. *Synthetic critical storm peak 24-hour pattern*

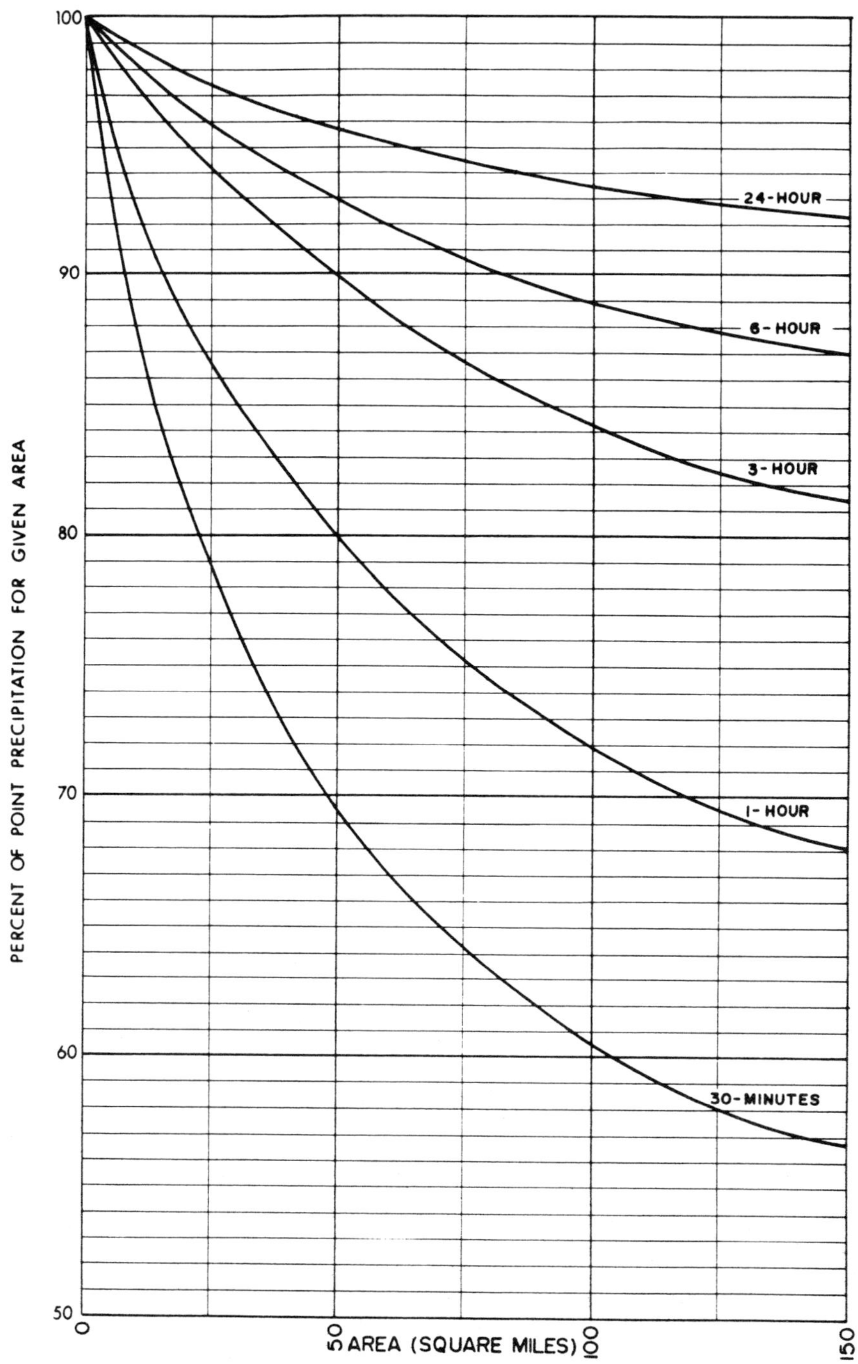

Fig. 5.11. NOAA Atlas 2 depth-area curves

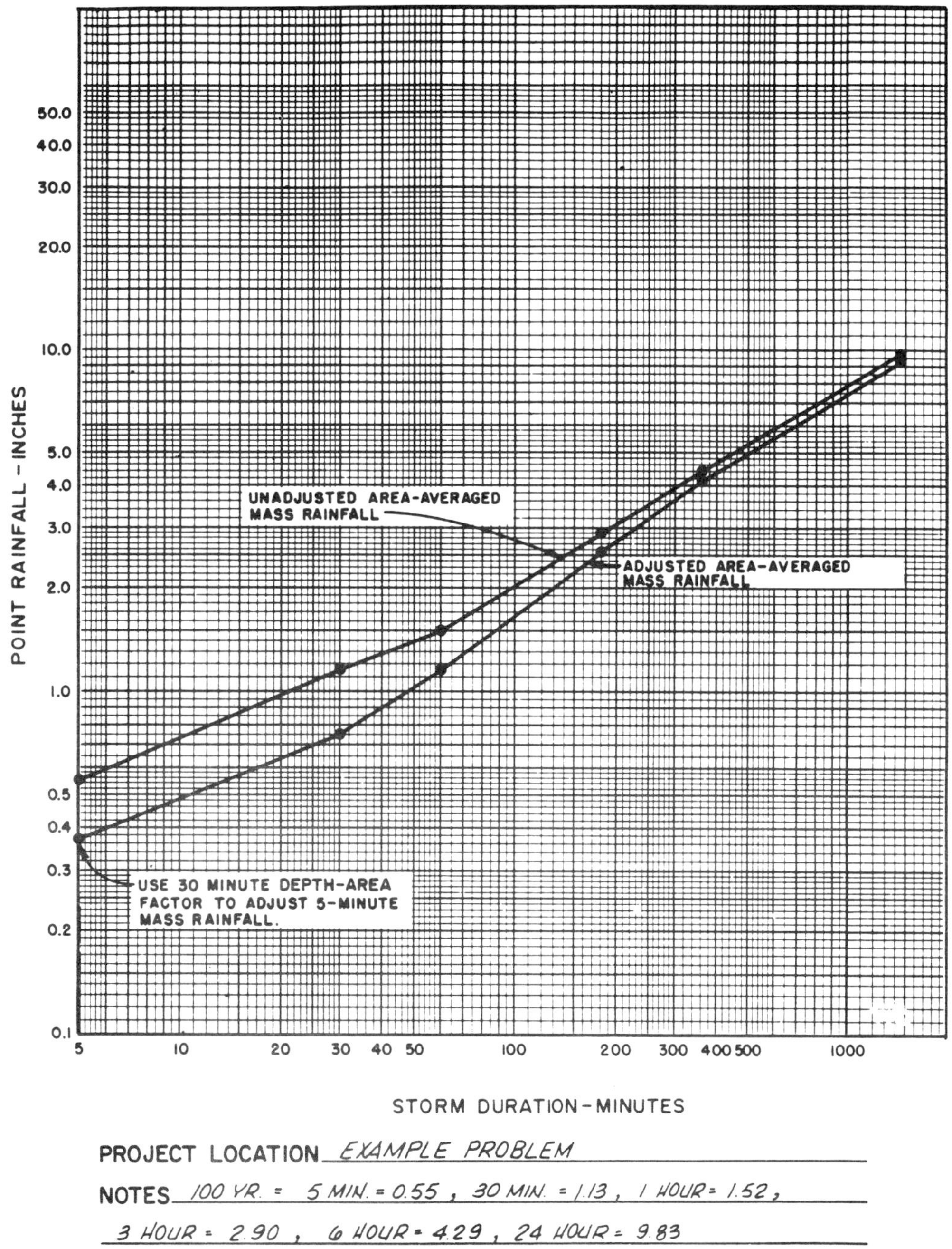

Fig. 5.12. Area-averaged mass rainfall plotting sheet

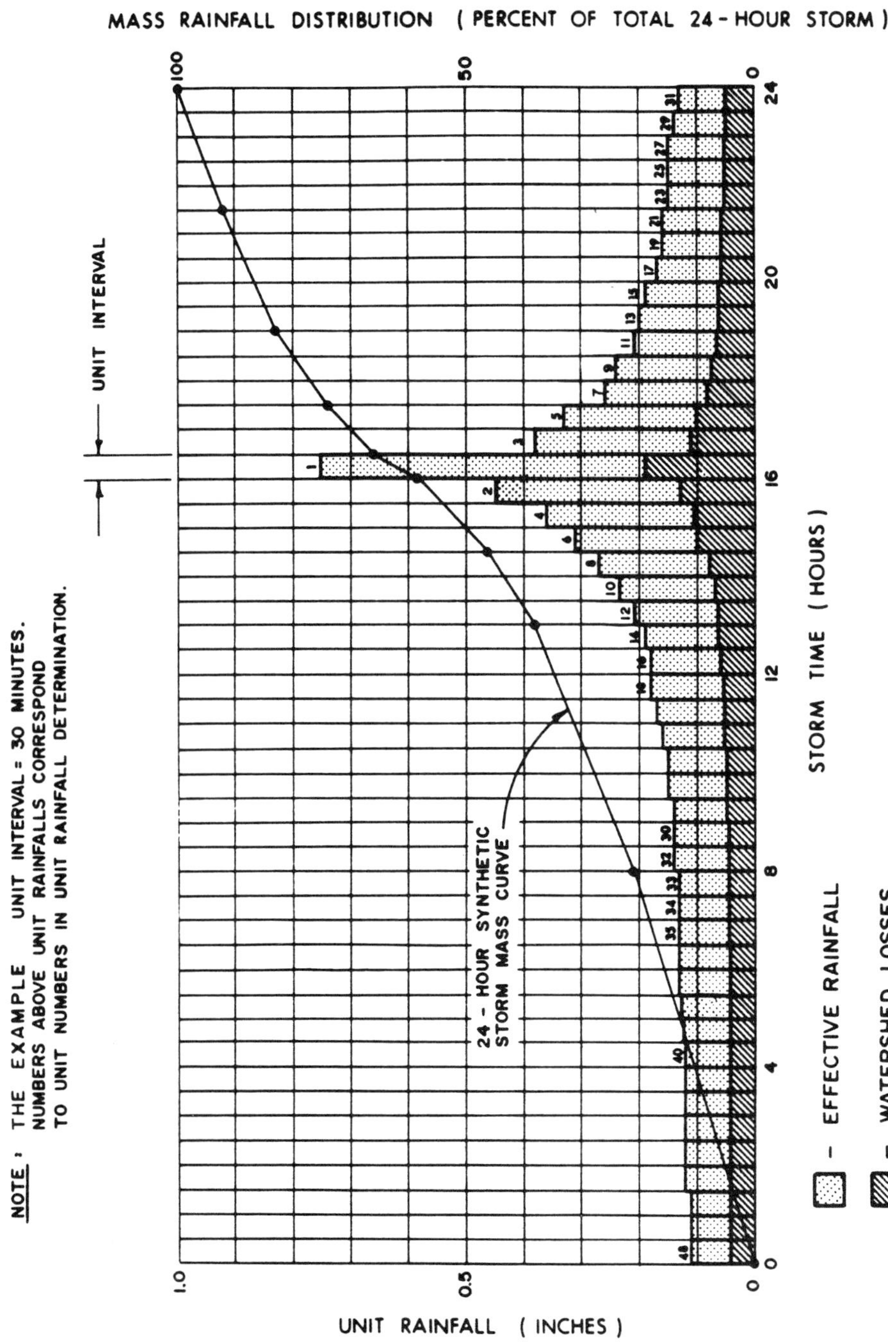

Fig. 5.13. Example synthetic 24-hour critical storm

Design Storm Point Precipitation

In order to generate a runoff hydrograph, a design storm rainfall pattern needs to be developed which incorporates the local rainfall intensity-duration tendencies. Generally, point precipitation isohyetal information is adequate (e.g. NOAA Atlas 2, 1973) for estimating local rainfall depths of a desired return frequency. Such isohyetal maps can be used to obtain peak point precipitation depths for storm durations from 15 minutes to 24 hours and with return frequencies from 2 years to 100 years. Because additional rainfall data may alter the statistical patterns used to develop the isohyetal maps, such generalized data should always be verified with a recent frequency analysis for all rain gages within the vicinity of the study watershed.

The peak 5-minute, 30-minute, 1-hour, 3-hour, 6-hour, and 24-hour area-averaged point precipitation values should be determined to evaluate the critical design storm pattern used to develop the runoff hydrograph. Extrapolation procedures are often used to estimate intermediate duration point precipitation values. For example, it may be assumed that the peak 3-hour point precipitation value is related to the peak 1-hour and 6-hour point precipitation values by straightline interpolation on log-log paper. Using the appropriate 1-hour point precipitation value, the peak 5-minute and 30-minute point rainfalls can be estimated using Fig. 5.6.

To account for the areal effects on point precipitation values, reduction factors can be used to adjust the area-averaged point rainfalls (e.g. NOAA Atlas 2, or locally determined depth-area curves)。 In the unit hydrograph approach, the volume of watershed runoff is related to the watershed average depth of precipitation rather than the depth at specific points within the watershed. Depth-area curves approximately relate the average of all point precipitation values for a specific duration and frequency to the average depth of precipitation within the watershed for the same duration and frequency. The NOAA Atlas depth-area curves of Fig. 5.11 provide adjustment factors for the 30-minute, 1-hour, 3-hour, 6-hour, and 24-hour point precipitation values. These curves are an average for the entire United States, however, and local depth-area tendencies may differ substantially.

S.C.S. Hydrologic Soil Groups

A major factor affecting infiltration is the condition of the soil. The soil surface characteristics, ability to transmit water through subsurface layers and the available soil storage capacity are all variables in defining the infiltration rate function. The S.C.S. classified more than 4000 soil types into four general categories which provide a classification of infiltration rates and corresponding runoff rates. These soil groups are defined in Table 5.4. The S.C.S. also uses curve numbers (CN) to describe the soil runoff potential. Further studies in California watersheds, however, provide some adjustments to the CN values. Runoff index numbers (RI) are used in this chapter to describe the use of these modified CN values.

Soil infiltration rates can be estimated for each of the soil groups by laboratory studies and measurements. Such measurements show that given a sufficient water supply, an initially dry soil will have an associated infiltration rate which essentially decreases with time. Consequently, if the soil is subjected to a continuous rainfall rate which exceeds the infiltration capacity, the infiltration loss rate will decrease with increasing storm duration. Depending on the condition and cover of the soil, a relative minimum infiltration rate for the S.C.S. soil groups are given in Table 5.5 (e.g. U.S. Bureau of Reclamation, 1973).

TABLE 5.4. S.C.S. HYDROLOGIC SOIL GROUPS

(A): Low runoff potential. Soils having a high infiltration rate even when thoroughly wetted. Consists chiefly of deep, well drained gravels and sands.

(B): Soils having moderate infiltration rates when thoroughly wetted. Consists mainly of moderately deep, well drained soils with moderately fine to moderately coarse texture.

(C): Soils having low infiltration rates when thoroughly wetted and consisting chiefly of soils with a layer that impedes the downward migration of water, or soils with moderately fine to fine texture.

(D): High runoff potential. Soils having a very low rate of infiltration when thoroughly wetted and consisting chiefly of clay soils with a high swelling potential, soils with a high water table (eliminating soil water storage capacity), soils with a claypan or clay layer at or near the soil surface, or shallow soils over a nearly impervious layer.

TABLE 5.5. MINIMUM INFILTRATION RATES

S.C.S. Soil Group	Infiltration Rate (in/hr)
A	0.30 - 0.45
B	0.15 - 0.30
C	0.05 - 0.15
D	0.00 - 0.10

Soil Cover

The type of vegetation or ground cover on a watershed, and the quality or density of that cover, have a major impact on the infiltration capacity of the soil. A widely used system for defining the effects of soil cover was developed by the S.C.S. and is based on the three categories given in Table 5.6.

In most cases, cover type and quality can be readily determined by an inspection of the watershed. U.S. Forest Service and S.C.S. personnel may also be helpful in estimation of such information. Cover types are listed in Fig. 5.14 which relate RI numbers for various cover qualities. General descriptions of cover types are given in Fig. 5.15.

TABLE 5.6. SOIL COVER QUALITY DEFINITIONS

POOR: Heavily grazed or regularly burned areas. Less than 50 percent of the ground surface is protected by plant cover or brush and tree canopy.

FAIR: Moderate cover with 50 percent to 75 percent of the ground surface protected.

GOOD: Heavy or dense cover with more than 75 percent of the ground surface protected.

Antecedent Moisture Condition

In order to accomodate the effects of prior precipitation events, the S.C.S. antecedent moisture condition (AMC) criteria are often used to adjust the infiltration loss rates. AMC curves were determined by analysis of rainfall-runoff data collected from a large number of watersheds where vegetative cover and soil conditions were known. For each watershed, the 24-hour runoff volume was plotted against the 24-hour rainfall. To accomodate scatter in the data, three runoff index numbers (that is, S.C.S. CN values) were associated to each watershed. The RI number which resulted in an equal number of high and low runoff predictions was defined to be an AMC II condition. Lower and upper enveloping RI numbers were associated to AMC I and AMC III conditions, respectively. The AMC designations are defined as follows:

TABLE 5.7. ANTECEDENT MOISTURE CONDITION (AMC) DEFINITIONS

AMC I: Lowest runoff potential. The watershed soils are dry enough to allow satisfactory grading or cultivation to take place.

AMC II: Moderate runoff potential. An intermediate condition which is generally assumed for annual flood studies.

AMC III: Highest runoff potential. The watershed is practically saturated from prior precipitation events. Heavy rainfall (or light rainfall with low temperatures) has occurred within the last five days.

Runoff Index Numbers of Hydrologic Soil-Cover Complexes For Pervious Areas-AMC II					
Cover Type	Quality of Cover (2)	Soil Group			
		A	B	C	D
NATURAL COVERS -					
Barren (Rockland, eroded and graded land)		78	86	91	93
Chaparrel, Broadleaf (Manzonita, ceanothus and scrub oak)	Poor Fair Good	53 40 31	70 63 57	80 75 71	85 81 78
Chaparrel, Narrowleaf (Chamise and redshank)	Poor Fair	71 55	82 72	88 81	91 86
Grass, Annual or Perennial	Poor Fair Good	67 50 38	78 69 61	86 79 74	89 84 80
Meadows or Cienegas (Areas with seasonally high water table, principal vegetation is sod forming grass)	Poor Fair Good	63 51 30	77 70 58	85 80 71	88 84 78
Open Brush (Soft wood shrubs - buckwheat, sage, etc.)	Poor Fair Good	62 46 41	76 66 63	84 77 75	88 83 81
Woodland (Coniferous or broadleaf trees predominate. Canopy density is at least 50 percent.)	Poor Fair Good	45 36 25	66 60 55	77 73 70	83 79 77
Woodland, Grass (Coniferous or broadleaf trees with canopy density from 20 to 50 percent)	Poor Fair Good	57 44 33	73 65 58	82 77 72	86 82 79
URBAN COVERS -					
Residential or Commercial Landscaping (Lawn, shrubs, etc.)	Good	32	56	69	75
Turf (Irrigated and mowed grass)	Poor Fair Good	58 44 33	74 65 58	83 77 72	87 82 79
AGRICULTURAL COVERS -					
Fallow (Land plowed but not tilled or seeded)		77	86	91	94

Fig. 5.14. (1 of 2) Runoff index numbers for pervious areas

Cover Type	Quality of Cover (2)	Soil Group A	B	C	D
Runoff Index Numbers of Hydrologic Soil-Cover Complexes For Pervious Areas-AMC II					
AGRICULTURAL COVERS (Continued)					
Legumes, Close Seeded (Alfalfa, sweetclover, timothy, etc.)	Poor	66	77	85	89
	Good	58	72	81	85
Orchards, Evergreen (Citrus, avocados, etc.)	Poor	57	73	82	86
	Fair	44	65	77	82
	Good	33	58	72	79
Pasture, Dryland (Annual grasses)	Poor	68	79	86	89
	Fair	49	69	79	84
	Good	39	61	74	80
Pasture, Irrigated (Legumes and perennial grass)	Poor	58	74	83	87
	Fair	44	65	77	82
	Good	33	58	72	79
Row Crops (Field crops - tomatoes, sugar beets, etc.)	Poor	72	81	88	91
	Good	67	78	85	89
Small grain (Wheat, oats, barley, etc.)	Poor	65	76	84	88
	Good	63	75	83	87

Notes:

1. All runoff index (RI) numbers are for Antecedent Moisture Condition (AMC) II.

2. Quality of cover definitions:

   Poor-Heavily grazed or regularly burned areas. Less than 50 percent of the ground surface is protected by plant cover or brush and tree canopy.

   Fair-Moderate cover with 50 percent to 75 percent of the ground surface protected.

   Good-Heavy or dense cover with more than 75 percent of the ground surface protected.

Fig. 5.14. (2 of 2) Runoff index numbers for pervious areas

Residential Landscaping (Lawn, Shrubs, etc.) - The pervious portions of commercial establishments, single and multiple family dwellings, trailer parks and schools where the predominant land cover is lawn, shrubbery and trees.

Row Crops - Lettuce, tomatoes, beets, tulips or any field crop planted in rows far enough apart that most of the soil surface is exposed to rainfall impact throughout the growing season. At plowing, planting and harvest times it is equivalent to fallow.

Small Grain - Wheat, oats, barley, flax, etc. planted in rows close enough that the soil surface is not exposed except during planting and shortly thereafter.

Legumes - Alfalfa, sweetclover, timothy, etc. and combinations are either planted in close rows or broadcast.

Fallow - Fallow land is land plowed but not yet seeded or tilled.

Woodland - grass - Areas with an open cover of broadleaf or coniferous trees usually live oak and pines, with the intervening ground space occupied by annual grasses or weeds. The trees may occur singly or in small clumps. Canopy density, the amount of ground surface shaded at high noon, is from 20 to 50 percent.

Woodland - Areas on which coniferous or broadleaf trees predominate. The canopy density is at least 50 percent. Open areas may have a cover of annual or perennial grasses or of brush. Herbaceous plant cover under the trees is usually sparse because of leaf or needle litter accumulation.

Chaparral - Land on which the principal vegetation consists of evergreen shrubs with broad, hard, stiff leaves such as manzonita, ceanothus and scrub oak. The brush cover is usually dense or moderately dense. Diffusely branched evergreen shrubs with fine needle-like leaves, such as chamise and redchank, with dense high growth are also included in this soil cover.

Annual Grass - Land on which the principal vegetation consists of annual grasses and weeds such as annual bromes, wild barley, soft chess, ryegrass and filaree.

Irrigated Pasture - Irrigated land planted to perennial grasses and legumes for production of forage and which is cultivated only to establish or renew the stand of plants. Dry land pasture is considered as annual grass.

Meadow - Land areas with seasonally high water table, locally called cienegas. Principal vegetation consists of sod-forming grasses interspersed with other plants.

Orchard (Deciduous) - Land planted to such deciduous trees as apples, apricots, pears, walnuts, and almonds.

Orchard (Evergreen) - Land planted to evergreen trees which include citrus and avocados and coniferous plantings.

Turf - Golf courses, parks and similar lands where the predominant cover is irrigated mowed close-grown turf grass. Parks in which trees are dense may be classified as woodland.

Fig. 5.15. S.C.S. cover type descriptions

As a guide in selecting the watershed AMC condition for the hydrologic analysis, the following agricultural season precipitation limits may be considered:

TABLE 5.8. ANTECEDENT MOISTURE CONDITION CRITERIA

AMC	Total 5-Day Antecedent Rainfall (inch) (dormant season)	(growing season)
I	less than 0.5	less than 1.4
II	0.5 - 1.1	1.4 - 2.1
III	over 1.1	over 2.1

Should the RI number be modified for AMC I or III, the following table can be used to determine an adjusted RI number:

TABLE 5.9. AMC RUNOFF INDEX NUMBER RELATIONSHIPS

RI For AMC II	Corresponding RI for AMC Condition I	III
100	100	100
95	87	99
90	78	98
85	70	97
80	63	94
75	57	91
70	51	87
65	45	83
60	40	79
55	35	75
50	31	70
45	27	65
40	23	60
35	19	55
30	15	50
25	12	45
20	9	39
15	7	33
10	4	26
5	2	17
0	0	0

Impervious Areas

In analysis of urban watersheds, the effects of the impervious surfaces on the assumed area-averaged infiltration rate for the entire watershed must be included. Estimated ranges of impervious percentages for various types of land use development are given in Fig. 5.16. Values given are for the actual percentage of area covered by an impervious surface. However, the actual impervious area needs to be reduced due to the local drainage practices. For example, an impervious surface may drain across a pervious surface where infiltration may take place. To account for this infiltration, the actual impervious area may be reduced by a factor such as 10 percent. This type of adjustment is included in the estimation of watershed loss rates for use in the synthetic unit hydrograph method.

Watershed Development Conditions

Should the flood control facilities be expected to provide for the public protection for an extended period, then the maximum urbanization should be assumed in determining the watershed loss rates. All available long range urbanization master plans should be examined to insure that reasonable land use assumptions are included in the analysis. Particular attention should be directed to local landscape practices. For example, it is common to use ornamental gravels underlain by impervious plastic materials as a substitute for lawns and shrubs, resulting in an increase in the effective impervious percentage.

Estimation of Infiltration Rates

The S.C.S. approach assumes that daily storm rainfall is related to the subsequent runoff by

$$R = (P - Ia)^2/(P - Ia + S) \qquad (5.7)$$

where

R = total runoff (in.)
P = total precipitation (in.)
Ia = initial abstraction
$S = 1000/RI - 10$

Figure 5.17 provides a plot of equation (5.7) as a function of RI numbers. In the figure, it is assumed that the initial abstraction can be estimated by

$$Ia = 0.2S \qquad (5.8)$$

Other studies suggest that the above S.C.S. value of (5.8) is too high and a value of 0.05S to 0.10S is more appropriate (Aron et al., 1977). As discussed previously, severe design storm conditions may reduce the Ia value even further and may even be

ACTUAL IMPERVIOUS COVER		
Land Use (1)	Range-Percent	Recommended Value For Average Conditions-Percent (2)
Natural or Agriculture	0 - 10	0
Single Family Residential: (3)		
100,000 SF (2.5 acre) Lots	5 - 15	10
40,000 SF (1 acre) Lots	10 - 25	20
20,000 SF (1/2 acre) Lots	30 - 45	40
7,200 - 10,000 SF Lots	45 - 55	50
Multiple Family Residential:		
Condominiums	45 - 70	65
Apartments	65 - 90	80
Mobile Home Park	60 - 85	75
Commercial, Downtown Business or Industrial	80 - 100	90

Notes:

1. Land use should be based on ultimate development of the watershed. Long range master plans for the County and incorporated cities should be reviewed to insure reasonable land use assumptions.

2. Recommended values are based on average conditions which may not apply to a particular study area. The percentage impervious may vary greatly even on comparable sized lots due to differences in dwelling size, improvements, etc. Landscape practices should also be considered as it is common in some areas to use ornamental gravels underlain by impervious plastic materials in place of lawns and shrubs. A field investigation of a study area should always be made, and a review of aerial photos, where available, may assist in estimating the percentage of impervious cover in developed areas.

3. For typical horse ranch subdivisions increase impervious area 5 percent over the values recommended in the table above.

Fig. 5.16. Actual impervious cover for developed areas

entirely eliminated from the rainfall-runoff budget. If it is assumed that Ia is negligible, then (5.7) is simplified to

$$R = P^2/(P + S) \tag{5.9}$$

Letting P_e = P-Ia, both (5.7) and (5.9) can be written as

$$R = P_e^2/(P_e + S) \tag{5.10}$$

where P_e is the precipitation excess in inches.

To estimate infiltration rates from (5.10), the method proposed by Chen (1975) is illustrated by the example calculations shown in Table 5.10.

TABLE 5.10. EXAMPLE COMPUTATION OF INFILTRATION LOSSES

Storm Time (hrs.)	Total Rainfall (in.)	Total Runoff (in.)	Incremental Infiltration (in.)	Infiltration Rate (in/hr)
6.0	2.5	0.46	-	-
6.25	2.60	0.50	0.06	0.24
6.50	2.70	0.55	0.05	0.20
6.75	2.85	0.63	0.07	0.28
7.00	3.00	0.71	0.07	0.28
7.25	3.20	0.83	0.08	0.32
7.50	3.55	1.04	0.12	0.48
7.75	4.05	1.36	0.18	0.72
8.00	4.80	1.89	0.22	0.88

notes:

(i) storm represents the peak 2 hours of a severe historical storm in Southern California

(ii) losses estimated assuming Ia=0.2S, and RI=70 for the pervious portions of the watershed

The incremental computation of the infiltration losses in Table 5.10 is by means of scanning the appropriate rainfall-runoff curve from Fig. 5.17 with respect to incremental precipitation. A difficulty with this procedure is that unacceptably high (and low) infiltration loss rates may be predicted. Given a maximum and minimum infiltration rate, however, the unacceptable incremental infiltration losses can be adjusted and the total infiltration losses distributed throughout the remaining design storm. Rallison (1980) gives a detailed account of the origin of the S.C.S. rainfall-runoff equation. The runoff curve numbers (and the analogous runoff index numbers) were developed by relating daily (24- hour periods) runoff to precipitation records. It is noted

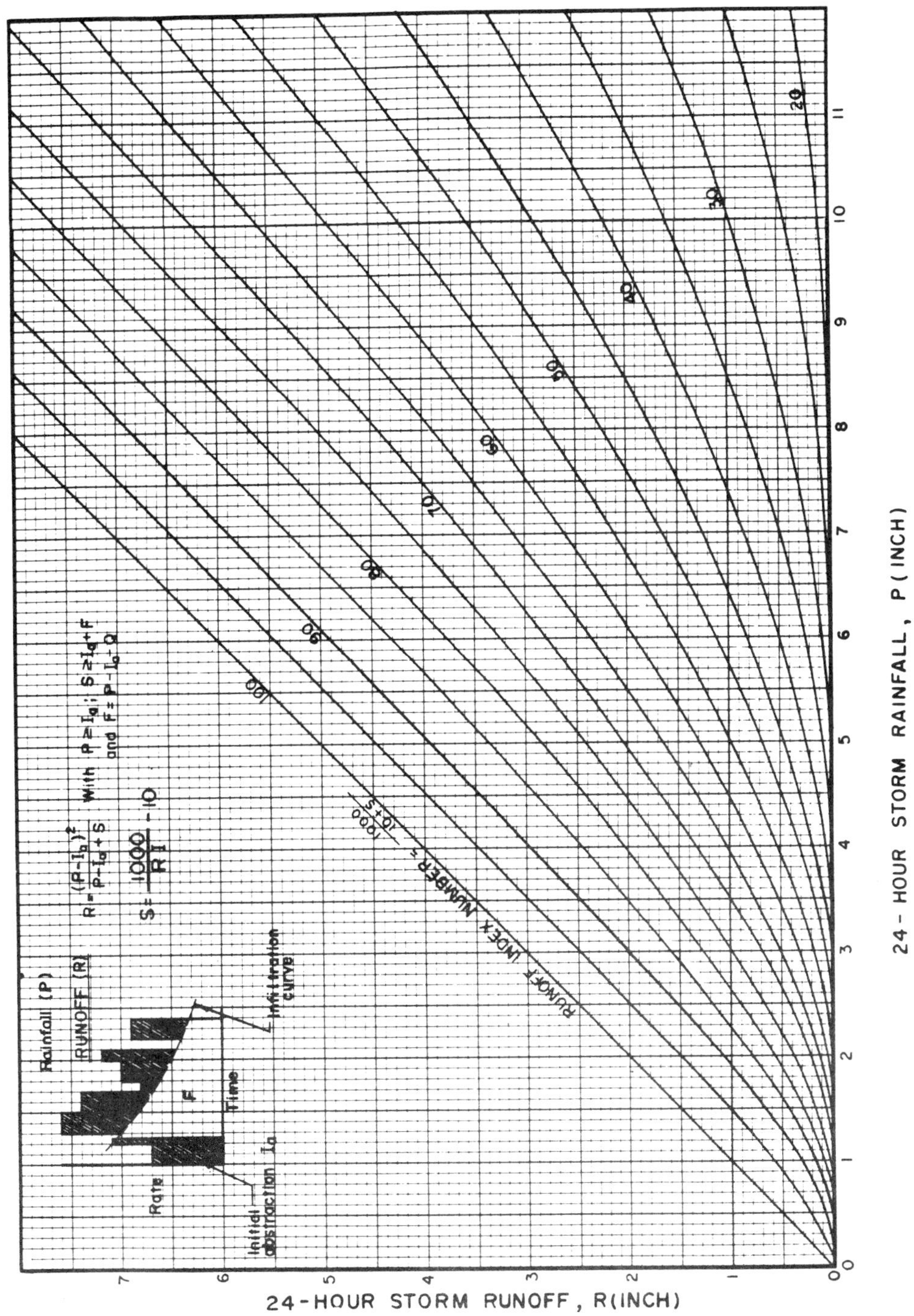

Fig. 5.17. *S.C.S. 24-hour storm rainfall-runoff relationships*

that these curves were not designed to be a physical definition of the incremental runoff estimation approach.

In estimating soil infiltration rates, an index of runoff potential is determined for each soil cover complex within the watershed. The RI scale has a range of 0 to 100, where a low RI number indicates a low runoff potential and a high RI number indicates a high runoff potential (an RI of 100 indicates total runoff). Selection of the RI number takes into account several major factors affecting the infiltration rate on pervious surfaces including hydrologic soils group, cover type and quality, and the antecedent moisture condition.

One approach for estimating a maximum limiting value for infiltration losses during the design storm event is utilized by several Southern Californian flood control agencies and is described by the following application problem. A severe design storm condition for many Californian watersheds would be 6 inches of precipitation within a duration of 6 hours. Additionally, the watershed design condition would be such that the initial abstraction Ia term is negligible due to significant prior storms and equation (5.9) would apply. For this situation, (5.9) reduces to

$$R = 6^2/(6 + S) \tag{5.11}$$

Assuming that the infiltration rate equals the maximum allowable value throughout this 6 hour severe storm condition gives

$$F_p = (P - R)\text{in.}/(6\text{ hrs}) \tag{5.12}$$

where F_p is the infiltration loss rate for the pervious areas. Rearranging terms results in an estimate for F_p in terms of the pervious area fraction RI number

$$F_p = (1000 - 10RI)/(1000 - 4RI) \tag{5.13}$$

Because this loss rate is for the pervious area only, a composite loss rate needs to be determined which represents both the pervious and impervious surfaces in the watershed. Adjustment of the loss rate for impervious surfaces can be made by

$$F = F_p(1.00 - 0.9A_i) \tag{5.14}$$

where

F = adjusted loss rate (in./hr.)

F_p = loss rate for pervious areas (in./hr.) derived from (5.13)

A_i = impervious area decimal fraction (Fig. 5.16)

The limiting infiltration rate from (5.14) may be used as the upper bound loss rate whenever it is exceeded by the incremental loss rate. Otherwise, the incremental loss rates are used with a lower bound infiltration rate estimated from Table 5.5. A second technique is to define a low loss rate percentage of rainfall to be attributed to infiltration losses during the lower intensity portion of the design storm event.

Assuming that the initial abstraction term (Ia) is negligible, then (5.9) may be used to define

$$Y^* = 1 - P/(P + S) \tag{5.15}$$

where

Y^* = low loss rate percentage (decimal fraction)
P = 24 hour precipitation (in.)
S = $(1000/RI^*) - 10$
RI^* = area-averaged composite runoff index number

To calculate the area-averaged runoff index used in (5.15), composite runoff index numbers must be computed for each RI number defined within the watershed by

$$RI_j^* = 100\,A_i + (RI_j)A_p \tag{5.16}$$

where

RI_j^* = composite runoff index number for area j
A_i = fraction of area j that is impervious
A_p = fraction of area j that is pervious

The above formula assumes an RI number of 100 (total runoff) for the impervious area fraction and uses an RI number selected from Fig. 5.14 for the pervious area fraction. The resulting composite runoff index number approximates the S.C.S. curve numbers for the subject development type, cover, perviousness, and soil group. The empirical relation of (5.16) is a simplification of a more complex relationship and, consequently, the RI associated to the impervious area fraction should not be reduced for the effects of effective imperviousness.

From the low loss rate percentage, the corresponding low loss rate, F', is given by

$$F' = Y^* I \tag{5.17}$$

where I is the rainfall intensity. It should be noted that the adjusted loss rate of (5.14) is an upper bound to the low loss rate of F'.

In the determination of the low loss rate percentage, the area-averaged 24- hour point rainfall (P) is computed and an appropriate composite runoff index number estimated, and both values are substituted into (5.15).

A third method in using a maximum (limiting) infiltration rate is incorporated into the Φ index approach which holds the loss rate as constant throughout the entire severe design storm event. Table 5.11 compares the three discussed methods in estimating the design storm watershed losses for the application problem presented in Table 5.10. In Table 5.11, it is assumed that the limiting F=0.44 in/hr and Y^*=0.40.

Comparing Tables 5.10 and 5.11 it is noted that limiting the incremental loss rate resulted in an additional 0.19 inches of runoff during the storm's peak 45 minutes of rainfall.

TABLE 5.11. COMPARISON OF INFILTRATION LOSS METHODS

Storm Time (hrs.)	Incremental Rainfall (in.)	Incremental Loss Approach (in.)	Low Loss Approach (in.)	Φ index Loss (in.)
6.0	0.00	-	-	-
6.25	0.10	0.06	0.04	0.10
6.50	0.10	0.05	0.04	0.10
6.75	0.15	0.07	0.06	0.11
7.00	0.15	0.07	0.06	0.11
7.25	0.20	0.08	0.08	0.11
7.50	0.35	0.11	0.11	0.11
7.75	0.50	0.11	0.11	0.11
8.00	0.75	0.11	0.11	0.11
sum:	(2.30)	(0.66)	(0.61)	(0.86)

Synthetic Runoff Hydrograph Development

The several hydrologic elements of the critical storm pattern, point rainfall determination, watershed lag estimation, and the unit hydrograph approximation are combined in the convolution of effective rainfall with the unit hydrograph. Figure 5.18 illustrates the procedure for developing a synthetic runoff hydrograph. Each unit interval (or unit period) of rainfall from the design storm pattern is split into the effective unit rainfall and the corresponding watershed unit loss. The resulting effective rainfall storm pattern is then combined with the assumed watershed unit hydrograph by a process called convolution to produce the synthetic runoff hydrograph. Figure 5.18 shows a 12 unit period storm pattern and the corresponding unit period runoff hydrographs. Summing together the 12 unit period runoff hydrographs results in the design storm runoff hydrograph.

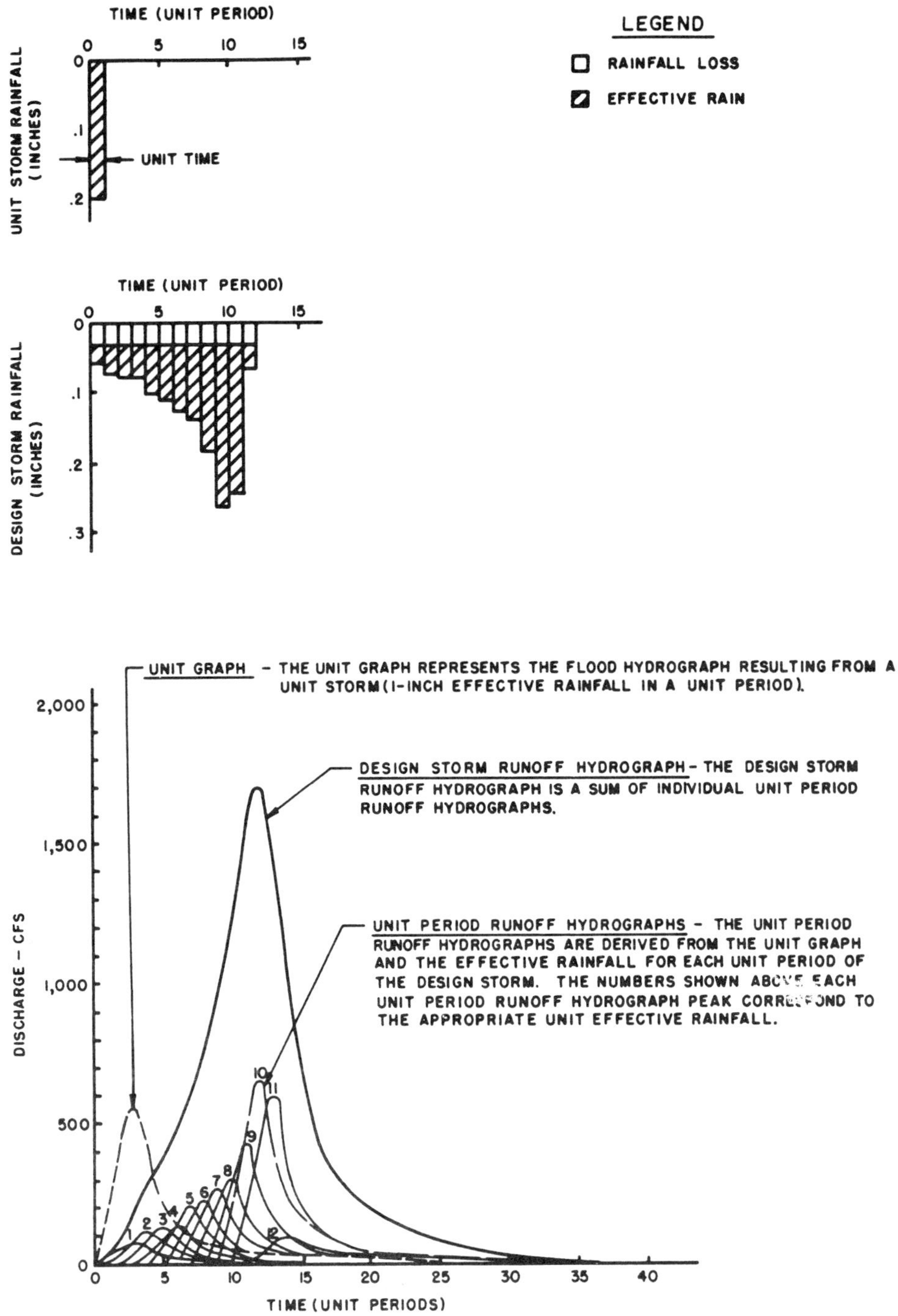

Fig. 5.18. Derivation of a runoff hydrograph

It should be noted that in the convolution approach, each unit period runoff hydrograph has the same base period regardless of the magnitude of the associated unit rainfall. Also, each unit period runoff hydrograph has a constant time to peak duration. Each of the unit period runoff hydrographs are seen to be directly related to the volume of the associated effective rainfalls. Additionally, in the summation process it is assumed that there are negligible restrictions to the watershed runoff flow rates which are represented by the runoff hydrograph. These restrictions would occur due to channel capacity problems and watershed detention effects. Each of these concerns should be addressed after the watershed runoff hydrograph is generated in order to ascertain whether the computed results are reasonable for the watershed conditions being studied. In many study situations, a single runoff hydrograph generation would be inappropriate due to the watershed conditions, and a more complex watershed model is required which includes submodels for channel and basin routing effects. Such a watershed model suitable for microcomputers is given in Hromadka (1983).

Other contributions to the runoff hydrograph such as baseflow and snowmelt can be summed to the synthetic runoff hydrograph. Generally, such considerations are neglected in the study of urban watersheds. However, detailed discussions of ancillary hydrologic processes are contained in texts such as Hjelmfelt and Cassidy (1975).

Instructions for a Synthetic Runoff Hydrograph Development

In order to describe the salient features of the supplied FORTRAN computer program, a process step chart is provided which describes each operation performed by the program (noted by a superscript star), and information which is to be provided by the program user. The computer program uses watershed S-curves developed for Southern California (U.S. Army Corps of Engineers) and also includes the dimensionless curvilinear S.C.S. unit hydrograph in S-graph form. Additionally, the program uses the NOAA Atlas 2 depth-area relationships of Fig. 5.11. This type of information can be easily replaced with local watershed S-curve information and other depth-area relationships in order to develop a runoff hydrograph generator program which represents local hydrologic characteristics. The basic elements of the program are as follows:

(1). The critical storm pattern is a composite 24 hour duration storm pattern composed of nested point rainfall peak intensities corresponding to each storm duration and assumed return frequency. The point rainfall values are adjusted by the depth-area relationships of the NOAA Atlas 2.

(2). Unit hydrographs are developed from the Southern California S-graphs determined by the U.S. Army Corps of Engineers. Also included is the S.C.S. curvilinear unit hydrograph. The time base of the unit hydrograph is determined by the watershed lag relationship of (5.1) and (5.2).

(3). Watershed losses are assumed equal to infiltration losses and are modeled by use of the adjusted loss rate of (5.14) and the low loss rate of (5.17).

Using the above three model assumptions, the synthetic runoff hydrograph is developed by the following steps:

I. Synthetic Unit Hydrograph Development

A. On a suitable topographic map of the watershed, determine the proposed drainage system boundaries.

B. From the watershed map, determine the following watershed physical factors and enter them on the computer information form of Fig. 5.19:

- A = drainage area (acres)
- L = length of the longest watercourse (feet)
- L_c = length along the longest watercourse, measured upstream to a point opposite the centroid of the watershed area (feet)
- H = elevation difference along watercourse (feet)

C*. Determine the watershed lag from (5.1) and (5.2) where the basin factor is selected from Fig. 5.2. The basin factor is entered on the form of Fig. 5.19.

D. Select a unit time period to be used for the modeling purposes. Usually, a unit period of 15 to 25 percent of the watershed lag time is adequate.

E*. Select a S-graph appropriate for the watershed (see Figs. 5.3 a,b,c,d, or Fig. 5.4). Determine the average percentage of the ultimate discharge for each unit period. In reading the percentage of ultimate discharge from the S-graph, the mean ordinate over the time period should be determined. The S-graphs may be averaged to approximate a combination of watershed runoff characteristics. Enter the desired S-graph proportions on the computer information form.

F*. Compute the unit distribution graph by subtracting successive unit percentages of ultimate discharge.

G*. Compute the ordinates of the synthetic unit hydrograph by multiplying the ordinates of the unit distribution graph by the ultimate discharge defined by

$$K(\text{cfs}) = 645A/T$$

where A is the drainage area in square miles, and T is the unit time period in hours.

PROJECT: ______________________ **DATE:**

____

ENGINEER: ______________________

1. Enter the length of the longest watercourse **(feet)** ________

2. Enter the length along the longest watershed watercourse measured from the point of concentration upstream to a point opposite the centroid of the watershed area **(feet)** ________

3. Enter the difference in elevation between the most remote point in the watershed and the point of concentration **(feet)** ________

4. Enter the basin factor ________

5. Enter the watershed area **(acres)** ________

6. Enter baseflow **(cfs/square mile)** ________

7. Enter S-Graph proportions **(decimal)**

   SCS[1] ________
   Valley ________
   Foothill ________
   Mountain ________
   Desert ________

8. Enter adjusted loss rate **(inch/hour)** ________

9. Enter low loss rate percentage **(decimal)** ________

10. Enter watershed area-averaged 5-minute point rainfall **(inches)*** ________

    Enter watershed area-averaged 30-minute point rainfall **(inches)*** ________

    Enter watershed area-averaged 1-hour point rainfall **(inches)*** ________

    Enter watershed area-averaged 3-hour point rainfall **(inches)*** ________

    Enter watershed area-averaged 6-hour point rainfall **(inches)*** ________

    Enter watershed area-averaged 24-hour point rainfall **(inches)*** ________

11. Enter 24-hour storm unit interval **(minutes)** ________

[1]Note: use either SCS or Valley, Foothill, Mountain, Desert
*Note: enter values unadjusted by depth-area factors

Fig. 5.19. Computer program data entry form

II. Synthetic Critical Storm Pattern Development

A. Develop point precipitation isohyetal maps for the 1 hour, 6-hour, and 24-hour durations, and calculate an area-averaged value for each of these durations.

B. Unless point precipitation data is available to determine area-averaged point precipitation values for the 5-minute, 30-minute, and 3-hour durations, use the area-averaged 1 hour value from step II.A to estimate the 5-minute and 30-minute values by plotting on Fig. 5.6 using a slope of the intensity-duration plot which is characteristic of the watershed rainfall records. The 3-hour value can be estimated from the 1-hour and 6-hour values by plotting on Fig. 5.20. Enter the area-averaged point rainfall values on the computer information form.

C*. Adjust all area-averaged point rainfall values for depth-area relationships such as given in the NOAA Atlas 2.

D*. Develop a synthetic critical storm mass rainfall plot such as shown in Fig. 5.12 using Fig. 5.20.

E*. Using the unit time period specified for the unit hydrograph determination, calculate the synthetic storm pattern unit rainfalls by successive subtractions of the mass rainfall curve values separated by a unit time period.

F*. Arrange the unit rainfalls into the critical storm pattern shown in Figs.5.10a,b,c. It should be noted that the computer stored rainfall pattern sequence places the peak 5-minute unit rainfall at storm time of 16 hours, and essentially distributes the remaining peak rainfalls symmetrically about the peak 5-minute value.

III. Flood Hydrograph Development

A. Find the pervious area fraction mean infiltration loss rate. Adjust the loss rate for the effective impervious area fraction by

$$F = F_p(1.00 - 0.9A_i)$$

where

F = adjusted loss rate (inch/hour)
F_p = loss rate for the pervious area fraction
A_i = impervious area fraction

B. Compute the low loss rate, F', from (5.17). Use the low loss rate for the watershed loss rate whenever the adjusted loss rate exceeds the low loss rate. Enter the adjusted loss rate and the low loss rate percentage of (5.15) on the computer information form.

C*. Compute the unit effective rainfalls by subtracting the unit losses from the unit rainfalls of the synthetic critical storm pattern.

D*. Compute the watershed runoff hydrograph (see Fig. 5.18):

1. Multiply the unit effective rainfall for the first time period by each unit value of the synthetic unit hydrograph to produce the runoff hydrograph corresponding to the unit period.
2. Repeat the above procedure for each unit effective rainfall, advancing the resulting unit period runoff hydrograph by one unit period.
3. Sum the unit period runoff hydrographs to produce the watershed runoff hydrograph.

E*. Add the baseflow to the flood hydrograph. Enter the baseflow on the computer information form.

A Synthetic Runoff Hydrograph Computer Program

The detailed description of procedures for developing a 24-hour storm runoff hydrograph can be directly formulated into a FORTRAN computer program suitable for currently available microcomputer systems. The following table lists the several subroutines used in the computer program:

TABLE 5.12. RUNOFF HYDROGRAPH PROGRAM SUBROUTINES

Program Number	Program Description
21	Main program for developing the effective rainfall storm pattern distribution and the resulting runoff hydrograph.
22 (no input)	Synthetic unit hydrograph generator which develops a unit hydrograph from the S-curves of Figs. 5.3a,b,c,d or the S.C.S. unit hydrograph of Fig. 5.4.
23 (no input)	Output plotting program which plots the runoff hydrograph flowrate and accumulated runoff volume as a function of storm time.

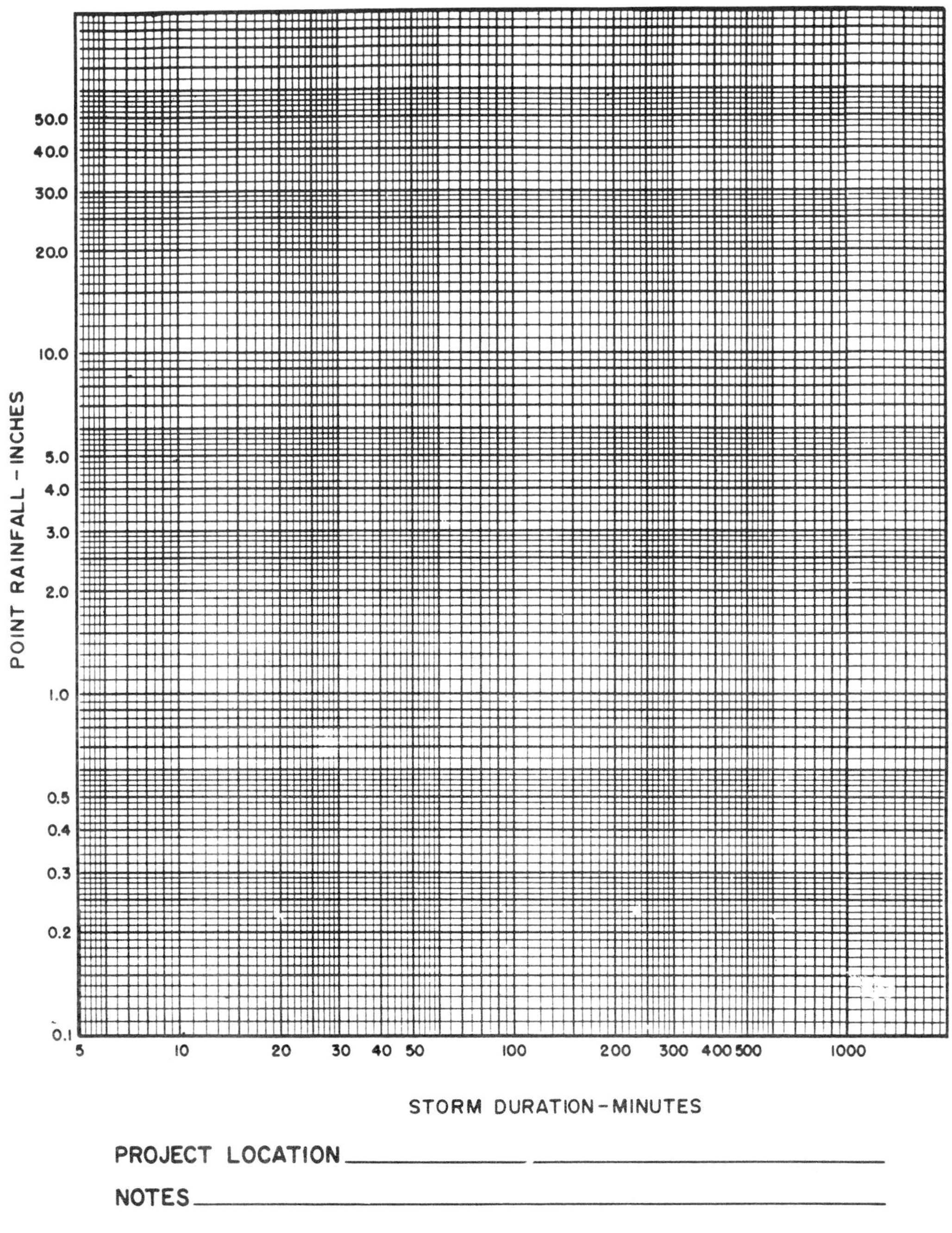

Fig. 5.20. Area-averaged mass rainfall plotting sheet

Chapter Bibliography

Aron, G., Miller, A. C., and Lakatos, D. F., "Infiltration Formula Based on S. C. S. Curve Number," ASCE Journal of Irrigation and Drainage., 103 (IRRA), (1977).

Chen, C. L. "Urban Storm Runoff Inlet Hydrograph Study," Rep. PRWG, 106-5, Utah Water Resources Lab., Utah State University, Logan, (1975).

Chen, C.L., "Infiltration Formulas by Curve Number Procedure," ASCE J. HYD. DIV., (108), (1982).

County of San Diego, "Hydrology Manual," San Diego, Ca., (1981).

Eagleson, P.S., **Dynamic Hydrology**, McGraw-Hill, (1970).

Espey, W. H., and Altman, D. G., "Nomographs for Ten-Minute Unit Hydrographs for Small Urban Watersheds," Rep. EPA-600/9-78-035 Environmental Protection Agency, Washington, D. C., (1978).

Federal Aviation Agency, "Dept. of Trans. Adv. Circular on Airport Drainage", Rep. A/C 050-5320-5B, Washington, D. C., (1970).

Freeze, R. A., "The Mechanism of Natural Groundwater Recharge and Discharge: 1. One-Dimensional, Vertical, Unsteady, Unsaturated Flow Above a Recharging or Discharging Groundwater Flow System," Wat. Res. Res., (5), (1969).

Garen, D. C., and Burges, S. J., "Approximate Error Bounds for Simulated Hydrographs," ASCE J. HYD. DIV., 107HY11, (1981).

Gray, D. D., et al, "Antecedent Moisture Condition Probabilities,"

ASCE J. Irr. Drainage Div., (108), (1982).

Green, W. H., and Ampt, G. A., "Studies on Soil Physics, I, the Flow of Air and Water Through Soils," J. Agr. Sci., (9), (1911).

Guymon, G. L., Harr, M. E., Berg, R. L., and Hromadka II, T. V., "A Probabilistic-Deterministic Analysis of One-Dimensional Ice Segregation in a Freezing Soil Column," Cold Regions Science and Technology, (5), (1981).

Helwig, O. J., Amorocho, J., and Finch, R.H., "Improvement of Nonlinear Rainfall-Runoff Model," ASCE J. HYD. DIV., 108HY7, (1982).

Hershfield, D. M., "Extreme Rainfall Relationships," ASCE J. HYD. DIV., 88 (HY6), (1962).

Hjelmfelt, A. T., "Curve-Number Procedure as Infiltration Method," ASCE J. HYD. DIV., (106), (1980).

Hjelmfelt, A. T., and Cassidy, J. J., **Hydrology for Engineers and Planners,** Iowa State University Press, (1975).

Holtan, H. N., "A Concept for Infiltration Estimates in Watershed Engineering," U. S. Dept. of Ag., ARS 41-51, (1961).

Horton, R. E., "The Role of Infiltration in the Hydrologic Cycle," Trans. A. G. U., (A), (1933).

Horton, R. E., "Determination of Infiltration Capacity for Large Drainage Basins," Trans. A. G. U., (18), 371, (1937).

Hromadka II, T. V., **Computer Methods in Urban Hydrology** , Lighthouse Publications, Mission Viejo, Ca., 1983.

Hromadka II, T. V., Nestlinger, A. J., and Schwarze, J. W., "A Modified S. C. S. Runoff Hydrograph Method," 10th Urban Drainage Symposium, Univ. of Kentucky, Lexington, Kentucky, (1983a).

Hromadka II, T. V., and Guymon, G. L., "San Bernardino County Hydrology Manual,", County of San Bernardino, Ca., (1983b).

Hromadka II, T. V., Clements, J. M., and Guymon, G. L., "Guidelines for Interactive Software in Water Resources Engineering," Wat. Res. Bul., Feb. (1983c).

Huff, F. A., "Time Distribution of Rainfall in Heavy Storms," Wat. Res. Res., (4), (1967).

Keifer, C. J., and Chu, H. H., "Synthetic Storm Pattern for Drainage Design," ASCE J. HYD. DIV., 83HY4, (1957).

Kibler, D. F., "Urban Stormwater Hydrology," A. G. U., Water Res, Mon., (7), (1982).

Kirpich, Z. P., "Time of Concentration of Small Agricultural Watersheds," Civ. Eng., 10(6), (1940).

Langbein, W. B., "Some Channel Storage and Unit Hydrograph Studies," Trans. A. G. U., (21), 620, (1940).

McCuen, R. H., and Bondelid, T. R., "Estimating Unit Hydrograph Peak Rate Factors," ASCE J. Irrig. and Drain., 109(2), (1983).

Mein, R. G., and Brown, B. M., "Sensitivity of Optimized Parameters in Watershed Models," Wat. Res. Res., (14), (1978).

Miller, J. F., Frederic, R. H., and Tracey, R. J., "Precipitation Frequency Atlas of the Conterminous Western United States," NOAA Atlas 2, Nat. Weather Serv., Silver Springs, Md., (1973).

Mockus, V. J., "National Engineering Handbook, Section 4, Hydrology," U. S. Dept. of Agri., S. C. S., Washington, D. C., (1972).

Myers, V. A., and Sehr, R. M., "A Methodology for Point-to-Area Rainfall Frequency Ratios," NOAA Tech. Rep. NWS 24, U. S. Dept. of Commerce, Washington, D. C., (1980).

Philip, J. R., "The Theory of Infiltration: 1. The Infiltration Equation and its Solution," Soil Sci., (83), (1957a).

Philip, J. R., "The Theory of Infiltration:2. The Profile at Infinity," Soil Sci. (83), (1957b).

Philip, J. R., "The Theory of Infiltration: 3. Moisture Profiles and Relation to Experiment," Soil Sci. (84), (1957d).

Philip, J. R., "The Theory of Infiltration: 4. Sorptivity and Algebraic Infiltration Equations," Soil Sci. (84), (1957d).

Philip, J. R., "The Theory of Infiltration: 5. The Influence of the Initial Moisture Content," Soil Sci. (84), (1957e).

Philip, J. R., "Evaporation and Moisture and Heat Fields in the Soil," J. Meteor. (14), (1957f).

Philip, J. R., "The Theory of Infiltration: 6. Effect of Water Depth over Soil," Soil Sci. (85), (1958a).

Pilgrim, D. H., and Cordery, I., "Rainfall Temporal Patterns for Design Floods," ASCE J. HYD. DIV., 101(HY1), (1975).

Rallison, R. E., "Origin and Evolution of the S. C. S. Runoff Equation," ASCE Watershed Management Symposium, Boise, Idaho, (1980).

Rubin, J., "Theoretical Analysis of Two-Dimensional Transient Flow of Water in Unsaturated and Partly Unsaturated Soil," Soil Sci. Soc. Amer. Proc., (32), (1968).

Rubin, J. and Steinhardt, R., "Soil Water Relations During Rain Infiltration: I. Theory," Soil Sci. Soc. Amer. Proc., (27), (1963).

Rubin, J., Steinhardt, R., and Reiniger, P., "Soil Water Relations During Rain Infiltration: II. Moisture Content Profiles During Rains of Low Intensities," Soil Sci. Soc. Amer. Proc., (28), (1964).

Sherman, L. K., "Streamflow from Rainfall by the Unit-Graph Method," Eng. News Rec., (108), (1932).

Snyder, F. F., "Synthetic Unit Hydrographs," EOS Trans. A. G. U., 19(1), (1938).

Terstriep, M. L., and Stall, J. B., "The Illinois Urban Drainage Area Simulator," ILLUDAS, Bul. 58, State Wat. Surv., Urbana, (1974).

U. S. Army Corps of Engineers, Los Angeles District, "Improved Procedure for Determining Drainage Area Lag Values," (1962).

U. S. Army Corps of Engineers, Los Angeles District, "Interim Report on Survey for Flood Control Tahquitz Creek, Ca.," (1963).

U. S. Dept. of Agriculture, Soil Conservation Service, "A Method for Estimating Volume and Rate of Runoff in Small Watersheds," Tech. Paper 149, Washington, D. C., (1973).

U. S. Dept. of Agriculture, Soil Conservation Service, "Engineering Field Manual for Conservation Practices," Notice CAL-1, EFM Exhibits CAL-2-11, and 2-12, (1969).

U. S. Dept. of Agriculture, Soil Conservation Service, "National Engineering Handbook," Section 4, Hydrology, Washington, D. C., (1969).

U. S. Dept. of Commerce, National Oceanic and Atmospheric Administration, National Weather Service, "NOAA Atlas 2, Precipitation-Frequency Atlas of the Western United States, Volume XI-California," (1973).

Wood, E. F., "An Analysis of the Effects of Parameter Uncertainty in Deterministic Hydrologic Models," Wat. Res. Res., (12), (1976).

Zaghloul, N. A., "SWMM Model and Level of Discretization," ASCE J. HYD. DIV., (107), (1981).

Hydrology Manuals:

Los Angeles County, 1971.
Orange County, 1973.
Riverside County, 1978.
San Diego County, 1981.
San Bernardino County, 1983.
State of Arizona Highway Dept., 1973.

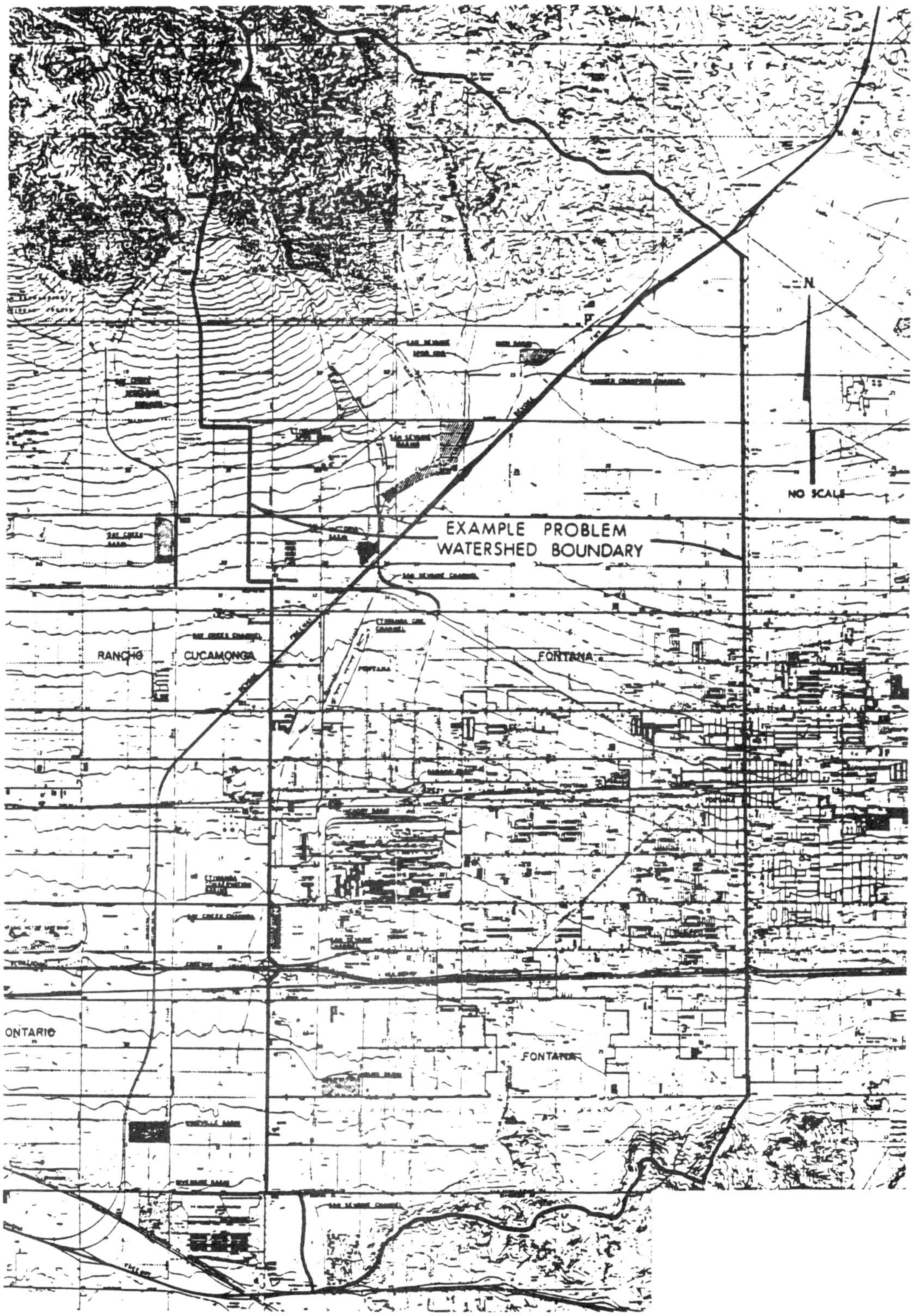

EXAMPLE PROBLEM WATERSHED

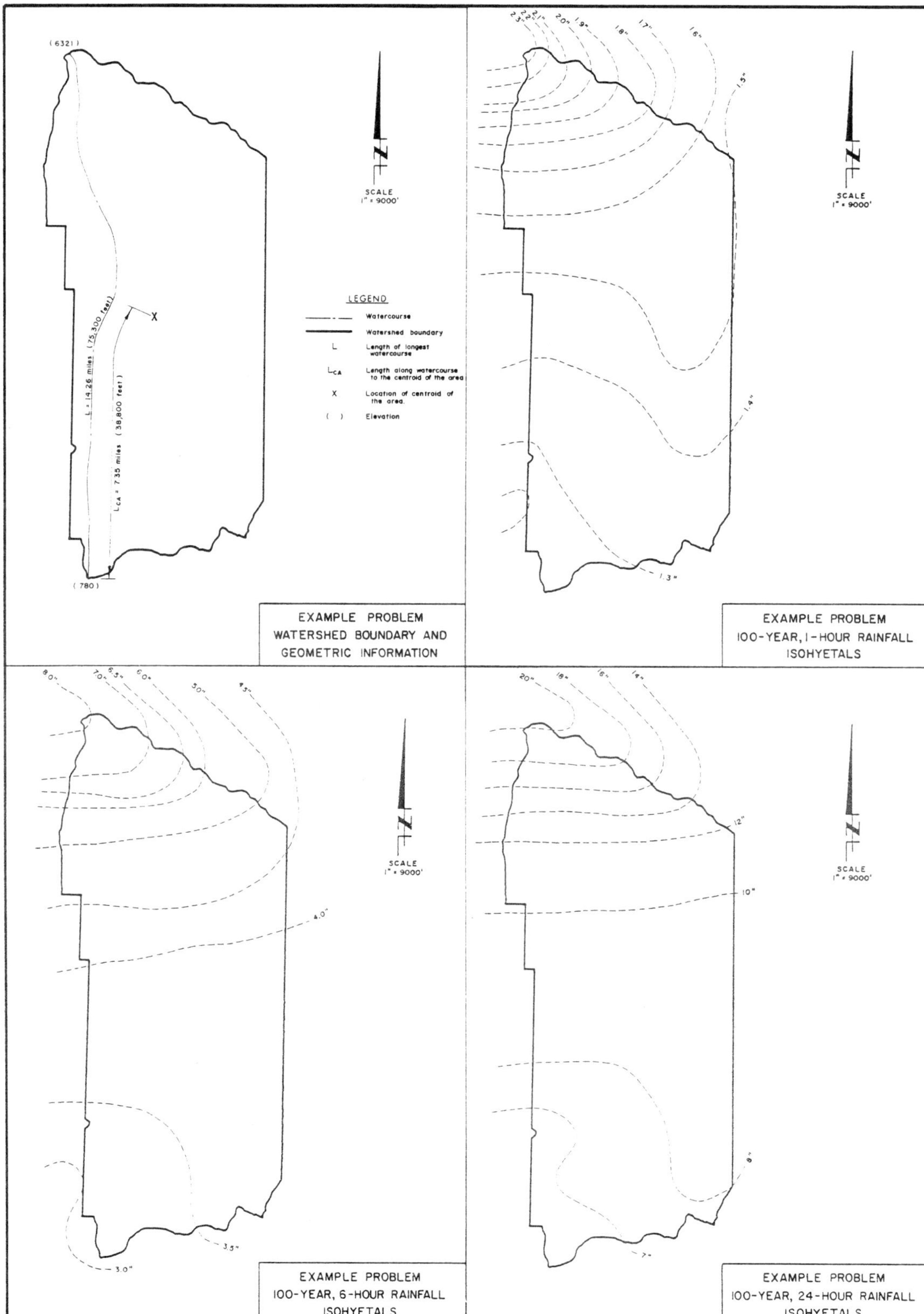

EXAMPLE PROBLEM
WATERSHED BOUNDARY AND
GEOMETRIC INFORMATION

EXAMPLE PROBLEM
100-YEAR, 1-HOUR RAINFALL
ISOHYETALS

EXAMPLE PROBLEM
100-YEAR, 6-HOUR RAINFALL
ISOHYETALS

EXAMPLE PROBLEM
100-YEAR, 24-HOUR RAINFALL
ISOHYETALS

PROJECT: EXAMPLE PROBLEM DATE:

ENGINEER:

1.	Enter the length of the longest watercourse **(feet)**	75,300
2.	Enter the length along the longest watershed watercourse measured from the point of concentration upstream to a point opposite the centroid of the watershed area **(feet)**	38,800
3.	Enter the difference in elevation between the most remote point in the watershed and the point of concentration **(feet)**	5,541
4.	Enter the basin factor	0.030
5.	Enter the watershed area **(acres)**	38,590
6.	Enter baseflow **(cfs/square mile)**	0.0
7.	Enter S-Graph proportions **(decimal)**	
	SCS[1]	0.0
	Valley	1.0
	Foothill	0.0
	Mountain	0.0
	Desert	0.0
8.	Enter adjusted loss rate **(inch/hour)**	0.30
9.	Enter low loss rate percentage **(decimal)**	0.90
10.	Enter watershed area-averaged 5-minute point rainfall **(inches)***	0.55
	Enter watershed area-averaged 30-minute point rainfall **(inches)***	1.13
	Enter watershed area-averaged 1-hour point rainfall **(inches)***	1.50
	Enter watershed area-averaged 3-hour point rainfall **(inches)***	2.90
	Enter watershed area-averaged 6-hour point rainfall **(inches)***	4.50
	Enter watershed area-averaged 24-hour point rainfall **(inches)***	9.80
11.	Enter 24-hour storm unit interval **(minutes)**	10.0

[1]Note: use either SCS or Valley, Foothill, Mountain, Desert
*Note: enter values unadjusted by depth-area factors

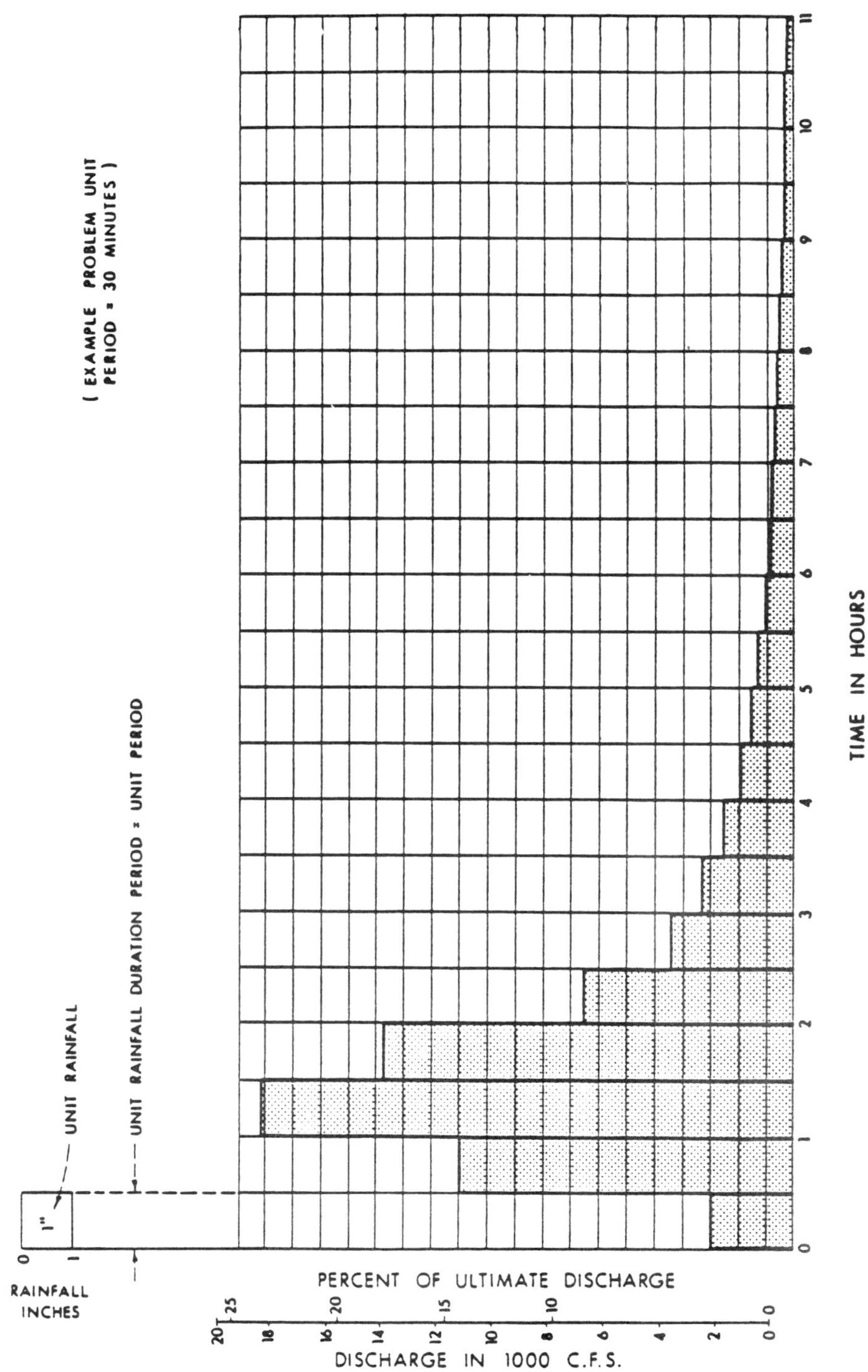

EXAMPLE UNIT DISTRIBUTION GRAPH

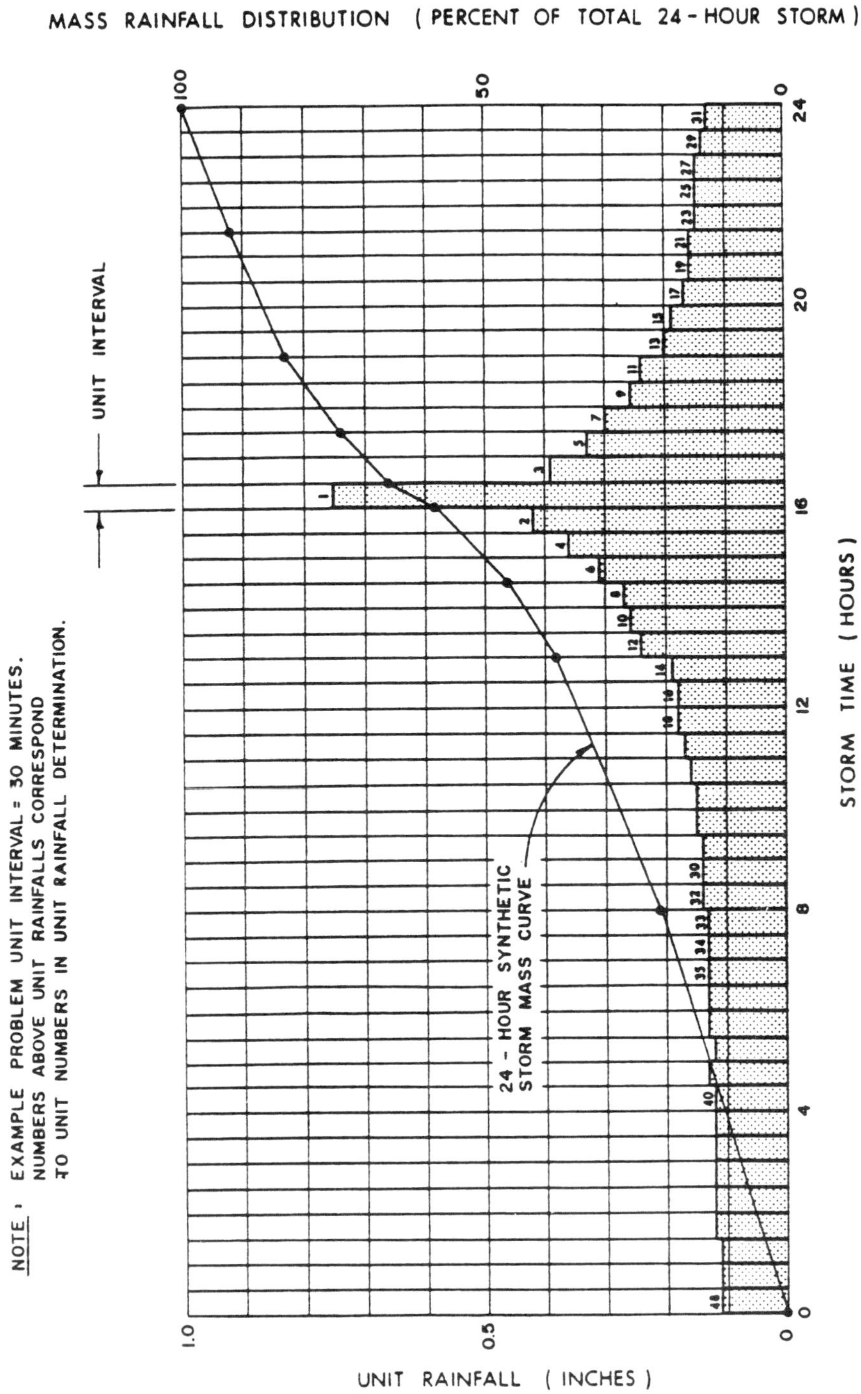

EXAMPLE PROBLEM SYNTHETIC 24 - HOUR CRITICAL STORM

```
**********DESCRIPTION OF RESULTS*********************************************
*                                                                           *
*                            EXAMPLE PROBLEM                                *
*                                                                           *
*****************************************************************************

*****************************************************************************
          WATERCOURSE LENGTH =   75300.000 FEET
          LENGTH FROM CONCENTRATION POINT TO CENTROID =   38800.000 FEET
          ELEVATION VARIATION ALONG WATERCOURSE =    5541.000 FEET
          MANNINGS FRICTION FACTOR ALONG WATERCOURSE =  .030
          WATERSHED AREA =  38590.000 ACRES
          WATERCOURSE "LAG" TIME =    1.359 HOURS
          UNIT HYDROGRAPH TIME UNIT =  10.000 MINUTES
          UNIT INTERVAL PERCENTAGE OF LAG-TIME =  12.268
          HYDROGRAPH BASEFLOW =    .000 CFS
          UNIFORM MEAN SOIL-LOSS(INCH/HOUR) =   .300
          LOW SOIL-LOSS RATE PERCENT(DECIMAL) =  .900
          VALLEY S-GRAPH SELECTED

          SPECIFIED PEAK 5-MINUTES RAINFALL(INCH)=   .55
          SPECIFIED PEAK 30-MINUTES RAINFALL(INCH) =  1.13
          SPECIFIED PEAK 1-HOUR RAINFALL(INCH) =  1.50
          SPECIFIED PEAK 3-HOUR RAINFALL(INCH) =  2.90
          SPECIFIED PEAK 6-HOUR RAINFALL(INCH) =  4.50
          SPECIFIED PEAK 24-HOUR RAINFALL(INCH)=  9.80

          *HYDROGRAPH MODEL #2 SPECIFIED*

          SAN BERNARDINO COUNTY AREA REDUCTION FACTORS:
            30-MINUTE FACTOR =  .665
            1-HOUR FACTOR =  .780
            3-HOUR FACTOR =  .884
            6-HOUR FACTOR =  .918
            24-HOUR FACTOR =  .948

<<<<<<<<<<<<<<<<<<<<<<<<<<<<<<<<<<<<<<>>>>>>>>>>>>>>>>>>>>>>>>>>>>>>>>>>>>>>>

                    Advanced Engineering Software [AES]
                              SERIAL No. M0008
                    REV. 1.2      RELEASE DATE: 7/22/82
```

UNIT HYDROGRAPH DETERMINATION

INTERVAL NUMBER	"S" GRAPH MEAN VALUES	UNIT HYDROGRAPH ORDINATES(CFS)
1	1.063	2481.010
2	3.437	5538.562
3	7.055	8444.191
4	12.251	12124.540
5	18.803	15288.240
6	26.768	18586.040
7	35.415	20178.950
8	44.247	20607.790
9	52.226	18618.550
10	58.433	14485.880
11	63.200	11123.090
12	67.039	8958.375
13	70.172	7309.129
14	72.659	5804.375
15	74.717	4801.316
16	76.542	4259.672
17	78.274	4041.979
18	79.762	3472.066
19	81.181	3311.589
20	82.422	2895.531
21	83.620	2795.193
22	84.710	2542.567
23	85.755	2440.056
24	86.656	2101.976
25	87.517	2009.827
26	88.269	1753.498
27	89.005	1718.568
28	89.741	1716.361
29	90.455	1667.331
30	91.023	1324.016
31	91.562	1258.679
32	92.102	1259.391
33	92.630	1233.398
34	93.130	1166.744
35	93.630	1166.744
36	94.130	1166.744
37	94.630	1166.744
38	95.130	1166.744
39	95.630	1166.744
40	96.130	1166.744
41	96.630	1166.744
42	97.130	1166.744
43	97.630	1166.744
44	98.130	1166.744
45	98.630	1166.744
46	99.130	1166.744
47	99.630	1166.744
48	100.000	862.560

UNIT PERIOD (NUMBER)	UNIT RAINFALL (INCHES)	UNIT SOIL-LOSS (INCHES)	EFFECTIVE RAINFALL (INCHES)
1	.0378	.0340	.0038
2	.0379	.0341	.0038
3	.0380	.0342	.0038
4	.0381	.0343	.0038
5	.0382	.0344	.0038
6	.0383	.0345	.0038
7	.0385	.0346	.0038
8	.0386	.0347	.0039
9	.0387	.0348	.0039
10	.0388	.0349	.0039
11	.0389	.0350	.0039
12	.0390	.0351	.0039
13	.0392	.0353	.0039
14	.0393	.0354	.0039
15	.0394	.0355	.0039
16	.0395	.0356	.0040
17	.0397	.0357	.0040
18	.0398	.0358	.0040
19	.0399	.0359	.0040
20	.0401	.0360	.0040
21	.0402	.0362	.0040
22	.0403	.0363	.0040
23	.0405	.0364	.0040
24	.0406	.0365	.0041
25	.0408	.0367	.0041
26	.0409	.0368	.0041
27	.0410	.0369	.0041
28	.0412	.0371	.0041
29	.0413	.0372	.0041
30	.0415	.0373	.0041
31	.0416	.0375	.0042
32	.0418	.0376	.0042
33	.0419	.0377	.0042
34	.0421	.0379	.0042
35	.0423	.0380	.0042
36	.0424	.0382	.0042
37	.0426	.0383	.0043
38	.0427	.0385	.0043
39	.0429	.0386	.0043
40	.0431	.0388	.0043
41	.0433	.0389	.0043
42	.0434	.0391	.0043
43	.0436	.0393	.0044
44	.0438	.0394	.0044
45	.0440	.0396	.0044
46	.0442	.0397	.0044
47	.0443	.0399	.0044
48	.0440	.0396	.0044
49	.0448	.0403	.0045
50	.0452	.0407	.0045
51	.0456	.0411	.0046
52	.0461	.0415	.0046
53	.0465	.0419	.0047
54	.0470	.0423	.0047
55	.0474	.0427	.0047
56	.0479	.0431	.0048
57	.0484	.0436	.0048
58	.0489	.0440	.0049

59	.0495	.0445	.0049
60	.0500	.0450	.0050
61	.0506	.0455	.0051
62	.0512	.0461	.0051
63	.0518	.0467	.0052
64	.0525	.0472	.0052
65	.0532	.0479	.0053
66	.0539	.0485	.0054
67	.0546	.0492	.0055
68	.0554	.0499	.0055
69	.0563	.0500	.0063
70	.0571	.0500	.0071
71	.0581	.0500	.0081
72	.0590	.0500	.0090
73	.0601	.0500	.0101
74	.0612	.0500	.0112
75	.0623	.0500	.0123
76	.0636	.0500	.0136
77	.0649	.0500	.0149
78	.0663	.0500	.0163
79	.0797	.0500	.0297
80	.0812	.0500	.0312
81	.0828	.0500	.0328
82	.0845	.0500	.0345
83	.0864	.0500	.0364
84	.0885	.0500	.0385
85	.0908	.0500	.0408
86	.0935	.0500	.0435
87	.0964	.0500	.0464
88	.1034	.0500	.0534
89	.1072	.0500	.0572
90	.1117	.0500	.0617
91	.1171	.0500	.0671
92	.1241	.0500	.0741
93	.1334	.0500	.0834
94	.1282	.0500	.0782
95	.1412	.0500	.0912
96	.1228	.0500	.0728
97	.4837	.0500	.4337
98	.1455	.0500	.0955
99	.1479	.0500	.0979
100	.1334	.0500	.0834
101	.1241	.0500	.0741
102	.1171	.0500	.0671
103	.1117	.0500	.0617
104	.1072	.0500	.0572
105	.1034	.0500	.0534
106	.0964	.0500	.0464
107	.0935	.0500	.0435
108	.0908	.0500	.0408
109	.0885	.0500	.0385
110	.0864	.0500	.0364
111	.0845	.0500	.0345
112	.0828	.0500	.0328
113	.0812	.0500	.0312
114	.0797	.0500	.0297
115	.0663	.0500	.0163
116	.0649	.0500	.0149
117	.0636	.0500	.0136
118	.0623	.0500	.0123
119	.0612	.0500	.0112
120	.0601	.0500	.0101
121	.0590	.0500	.0090
122	.0581	.0500	.0081
123	.0571	.0500	.0071
124	.0563	.0500	.0063

125	.0554	.0499	.0055
126	.0546	.0492	.0055
127	.0539	.0485	.0054
128	.0532	.0479	.0053
129	.0525	.0472	.0052
130	.0518	.0467	.0052
131	.0512	.0461	.0051
132	.0506	.0455	.0051
133	.0500	.0450	.0050
134	.0495	.0445	.0049
135	.0489	.0440	.0049
136	.0484	.0436	.0048
137	.0479	.0431	.0048
138	.0474	.0427	.0047
139	.0470	.0423	.0047
140	.0465	.0419	.0047
141	.0461	.0415	.0046
142	.0456	.0411	.0046
143	.0452	.0407	.0045
144	.0448	.0403	.0045

```
TOTAL SOIL-LOSS VOLUME(ACRE-FEET) =   20411.3600
TOTAL STORM RUNOFF VOLUME(ACRE-FEET) =    9456.9410
```

```
*******************************************************************************


*******************************************************************************

            HYDROGRAPH POINTS GIVEN IN FIVE-MINUTE INTERVALS

-------------------------------------------------------------------------------
INTERVAL# VOLUME(AF)   Q(CFS) 0.      5775.0    11550.0   17325.0    23100.0
-------------------------------------------------------------------------------
      1        .0645      9.37  Q          .         .         .          .
      2        .1291      9.37  Q          .         .         .          .
      3        .3379     30.32  Q          .         .         .          .
      4        .5467     30.32  Q          .         .         .          .
      5        .9757     62.30  Q          .         .         .          .
      6       1.4048     62.30  Q          .         .         .          .
      7       2.1505    108.28  Q          .         .         .          .
      8       2.8962    108.28  Q          .         .         .          .
      9       4.0418    166.34  Q          .         .         .          .
     10       5.1874    166.34  Q          .         .         .          .
     11       6.8198    237.03  Q          .         .         .          .
     12       8.4522    237.03  Q          .         .         .          .
     13      10.6143    313.94  Q          .         .         .          .
     14      12.7764    313.94  Q          .         .         .          .
     15      15.4810    392.70  Q          .         .         .          .
     16      18.1855    392.70  Q          .         .         .          .
     17      21.3823    464.18  Q          .         .         .          .
     18      24.5792    464.18  Q          .         .         .          .
     19      28.1623    520.27  Q          .         .         .          .
     20      31.7455    520.27  Q          .         .         .          .
     21      35.6287    563.84  Q          .         .         .          .
     22      39.5118    563.84  Q          .         .         .          .
     23      43.6397    599.36  VQ         .         .         .          .
     24      47.7675    599.36  VQ         .         .         .          .
     25      52.0979    628.77  VQ         .         .         .          .
     26      56.4283    628.77  VQ         .         .         .          .
     27      60.9228    652.60  VQ         .         .         .          .
     28      65.4173    652.60  VQ         .         .         .          .
     29      70.0504    672.73  VQ         .         .         .          .
     30      74.6835    672.73  VQ         .         .         .          .
     31      79.4416    690.88  VQ         .         .         .          .
     32      84.1997    690.88  VQ         .         .         .          .
     33      89.0776    708.27  VQ         .         .         .          .
     34      93.9555    708.27  VQ         .         .         .          .
     35      98.9389    723.59  VQ         .         .         .          .
     36     103.9223    723.59  VQ         .         .         .          .
     37     109.0074    738.36  VQ         .         .         .          .
     38     114.0925    738.36  VQ         .         .         .          .
     39     119.2688    751.61  VQ         .         .         .          .
     40     124.4452    751.61  VQ         .         .         .          .
     41     129.7106    764.54  VQ         .         .         .          .
     42     134.9761    764.54  VQ         .         .         .          .
     43     140.3244    776.58  VQ         .         .         .          .
     44     145.6728    776.58  VQ         .         .         .          .
     45     151.1017    788.29  VQ         .         .         .          .
     46     156.5307    788.29  VQ         .         .         .          .
     47     162.0318    798.77  VQ         .         .         .          .
     48     167.5330    798.77  VQ         .         .         .          .
     49     173.1043    808.95  VQ         .         .         .          .
     50     178.6755    808.95  VQ         .         .         .          .
     51     184.3106    818.22  VQ         .         .         .          .
```

```
184   1846.0400   8024.43   .      V  .  Q      .         .         .
185   1906.0800   8717.90   .       V .    Q    .         .         .
186   1966.1210   8717.90   .       V .    Q    .         .         .
187   2031.2960   9463.45   .       V .     Q   .         .         .
188   2096.4710   9463.45   .       V .     Q   .         .         .
189   2167.2440  10276.31   .        V.      Q  .         .         .
190   2238.0180  10276.31   .        V.      Q  .         .         .
191   2314.4120  11092.45   .        V.        Q.         .         .
192   2390.8060  11092.45   .         V        Q.         .         .
193   2478.6990  12762.11   .         V         . Q       .         .
194   2566.5930  12762.11   .         V         . Q       .         .
195   2667.4470  14644.00   .         .V        .    Q    .         .
196   2768.3010  14644.00   .         .V        .    Q    .         .
197   2881.5620  16445.54   .         . V       .       Q .         .
198   2994.8230  16445.54   .         . V       .       Q .         .
199   3121.9390  18457.18   .         .  V      .         .Q        .
200   3249.0540  18457.18   .         .  V      .         .Q        .
201   3387.9010  20160.57   .         .   V     .         .   Q     .
202   3526.7480  20160.57   .         .   V     .         .   Q     .
203   3676.7950  21786.85   .         .    V    .         .      Q  .
204   3826.8420  21786.85   .         .     V   .         .      Q  .
205   3983.0120  22675.99   .         .     V   .         .        Q.
206   4139.1800  22675.99   .         .      V  .         .        Q.
207   4297.6090  23004.44   .         .       V .         .        Q.
208   4456.0390  23004.44   .         .       V .         .        Q.
209   4609.8400  22332.24   .         .        V.         .       Q .
210   4763.6410  22332.24   .         .         V         .       Q .
211   4906.3010  20714.50   .         .         V         .    Q    .
212   5048.9610  20714.50   .         .          V        .    Q    .
213   5180.9180  19160.58   .         .         .V        .  Q      .
214   5312.8750  19160.58   .         .         . V       .  Q      .
215   5436.0160  17880.52   .         .         . V       Q         .
216   5559.1560  17880.52   .         .         .  V      Q         .
217   5674.3360  16724.19   .         .         .   V   Q .         .
218   5789.5160  16724.19   .         .         .   V   Q .         .
219   5896.9690  15602.48   .         .         .   V  Q  .         .
220   6004.4220  15602.48   .         .         .    V Q  .         .
221   6105.3630  14657.08   .         .         .    Q    .         .
222   6206.3050  14657.08   .         .         .    QV   .         .
223   6301.9060  13881.59   .         .         .   Q V   .         .
224   6397.5080  13881.59   .         .         .   Q  V  .         .
225   6488.8360  13261.05   .         .         . Q    V  .         .
226   6580.1640  13261.05   .         .         . Q    V  .         .
227   6666.6170  12553.45   .         .         .Q      V .         .
228   6753.0700  12553.45   .         .         .Q      V .         .
229   6835.7730  12008.55   .         .         Q       V .         .
230   6918.4770  12008.55   .         .         Q        V.         .
231   6996.7620  11367.36   .         .        Q.        V.         .
232   7075.0470  11367.36   .         .        Q.        V.         .
233   7149.6680  10835.00   .         .       Q .         V         .
234   7224.2890  10835.00   .         .       Q .         V         .
235   7294.7110  10225.30   .         .      Q  .         V         .
236   7365.1330  10225.30   .         .      Q  .          V        .
237   7431.6640   9660.36   .         .     Q   .         .V        .
238   7498.1950   9660.36   .         .     Q   .         .V        .
239   7560.0900   8987.30   .         .    Q    .         .V        .
240   7621.9840   8987.30   .         .    Q    .         . V       .
241   7679.8120   8397.21   .         .   Q     .         . V       .
242   7737.6410   8397.21   .         .   Q     .         . V       .
243   7791.0350   7753.07   .         .  Q      .         . V       .
244   7844.4300   7753.07   .         .  Q      .         .  V      .
245   7894.1910   7225.84   .         . Q       .         .  V      .
246   7943.9530   7225.84   .         . Q       .         .  V      .
247   7990.5740   6769.87   .         .Q        .         .  V      .
248   8037.1950   6769.87   .         .Q        .         .  V      .
```

PROGRAM 21: DATA ENTRY

```
Enter the length of the longest watercourse...(FEET) ===>     "XL"
:ALLOWABLE VALUES ARE [1000   ] TO [999999]

Enter the length along the longest watershed
watercourse measured from the point of
concentration upstream to a point opposite the
centroid of the watershed area................(FEET) ===>     "XLCA"
:ALLOWABLE VALUES ARE [250    ] TO [800000]

Enter the difference in elevation between
the most remote point in the watershed and
the point of concentration....................(FEET) ===>     "HH"
:ALLOWABLE VALUES ARE [.5     ] TO [100000]

------------------------------------------------------------------------------
    TYPE: EXIT to leave program ; TOP  to go to top of page
                                ; BACK to go back one page
```

```
BASIN       GENERAL DESCRIPTION OF DRAINAGE AREA
FACTOR      (Ref. U.S. Corps of Engineers)

N= .015 >>> Drainage area has fairly uniform gentle slopes.
            Most watercourses either improved or along paved streets.
N= .020 >>> Drainage has some graded and non-uniform gentle slopes.
            Over half of the area watercourses are improved or paved streets.
N= .025 >>> Drainage area is generally rolling with gentle side slopes.
            Some drainage improvements in the area - streets and canals.
N= .030 >>> Drainage area is generally rolling with rounded ridges and moderate
              side slopes.
            No drainage improvements exist in the area.
N= .040 >>> Drainage area is composed of steep upper canyon with moderate
              slopes in lower canyons.
            No drainage improvements exist in the area.
N= .050 >>> Drainage area is quite rugged with sharp edges and steep canyons.
            No drainage improvements exist in the area.
N= .200 >>> Drainage area has comparatively uniform slopes.
            No drainage improvements exist in the area.
    Enter basin factor (BETWEEN .005 AND .500).......... ===>     "XN"

------------------------------------------------------------------------------
    TYPE: EXIT to leave program ; TOP  to go to top of page
                                ; BACK to go back one page
```

```
Enter the watershed area(ACRES)..................... ===>    "A"
:ALLOWABLE VALUES ARE [10    ] TO [96000 ]


Enter baseflow(CFS/SQUARE-MILE)..................... ===>    "BASCON"
:ALLOWABLE VALUES ARE [0] TO [1000  ]









-------------------------------------------------------------------------------
    TYPE: EXIT to leave program ; TOP  to go to top of page
                                ; BACK to go back one page
```

```
UNIT-HYDROGRAPH "S" GRAPH OPTIONS:
    1: Valley Zone          2: Foothill Zone          3: Mountain Zone
    4: Desert Zone          5: Combination of options 1 through 4
    6: U.S. Soil Conservation Service (SCS)
Select appropriate S-GRAPH......................... ===>    "KODE1"

Enter percentage[DECIMAL NOTATION] of watershed
specified with VALLEY "S" curve.................... ===>    "PV"
:ALLOWABLE VALUES ARE [0.00  ] TO [1.00  ]

Enter percentage[DECIMAL NOTATION] of watershed
specified with FOOTHILL "S" curve.................. ===>    "PF"
:ALLOWABLE VALUES ARE [0.00  ] TO [1.00  ]

Enter percentage[DECIMAL NOTATION] of watershed
specified with MOUNTAIN "S" curve.................. ===>    "PM"
:ALLOWABLE VALUES ARE [0.00  ] TO [1.00  ]

DESERT "S" curve percentage[DECIMAL NOTATION] =         ["PD"]


-------------------------------------------------------------------------------
    TYPE: EXIT to leave program ; TOP  to go to top of page
                                ; BACK to go back one page
```

```
Enter adjusted loss rate(INCH/HOUR).................. ===>    "VSL"
:ALLOWABLE VALUES ARE [.01    ] TO [5.0    ]
{RECOMMENDED VALUES ARE [.20] TO [.40]}
NOTE: THE ADJUSTED LOSS RATE INCLUDES THE EFFECTS OF
IMPERVIOUSNESS, INITIAL ABSTRACTION, AND OTHER FACTORS.


Enter low loss rate percentage....[DECIMAL NOTATION] ===>    "SLP"
:ALLOWABLE VALUES ARE [.01    ] TO [.99    ]
NOTE: SUGGESTED VALUES ARE [.10] TO [.50]
NOTE: THE "LOW LOSS RATE PERCENTAGE" IS THE CONSTANT
PERCENTAGE OF RAINFALL TO BE USED FOR WATERSHED LOSSES.
THE MAXIMUM WATERSHED LOSS RATE IS 5.000(INCH/HOUR).




--------------------------------------------------------------------------------
TYPE: EXIT to leave program ; TOP  to go to top of page
                            ; BACK to go back one page
```

```
PRECIPITATION DEPTH-AREA FACTOR MODELS:

  1: NOAA Atlas II values (30-minute value used
     for 5-minute depth)

  2: User-specified depth-area factors

Select desired model................................ ===>    "DAOPT"




--------------------------------------------------------------------------------
TYPE: EXIT to leave program ; TOP  to go to top of page
                            ; BACK to go back one page
```

```
TO CONSTRUCT THE SYNTHETIC CRITICAL STORM PATTERN, AREA-AVERAGED
WATERSHED RAINFALL VALUES ARE NEEDED FOR THE PEAK 5-MINUTES,
30-MINUTES,1-,3-,6-,AND 24-HOURS OF RECORDED RAINFALL.

Enter watershed area-averaged  5-minute point
rainfall(INCHES).................................. ===>    "SEE BELOW"
:ALLOWABLE VALUES ARE [  .010] TO [    5.]

Enter  5-minute depth-area adjustment factor........ ===>    "SEE BELOW"
:ALLOWABLE VALUES ARE [  .010] TO [1.    ]

WATERSHED AREA-AVERAGED POINT RAINFALLS    DEPTH-AREA ADJUSTMENT FACTOR

      5-MINUTE   [ "R5"  ] INCHES               5-MINUTE   [ "FX5" ]
     30-MINUTE   [ "R30" ] INCHES              30-MINUTE   [ "FX30"]
        1-HOUR   [ "R1"  ] INCHES                 1-HOUR   [ "FX1" ]
        3-HOUR   [ "R3"  ] INCHES                 3-HOUR   [ "FX3" ]
        6-HOUR   [ "R6"  ] INCHES                 6-HOUR   [ "FX6" ]
       24-HOUR   [ "R24" ] INCHES                24-HOUR   [ "FX24"]

--------------------------------------------------------------------------------
TYPE: EXIT to leave program ; TOP  to go to top of page
                            ; BACK to go back one page
```

```
24-HOUR STORM UNIT-INTERVAL MODELS:

  MODEL     MODEL UNIT-INTERVAL
 NUMBER        (DURATION)
    1:          5-MINUTES        <=RECOMMENDED MODEL
    2:         10-MINUTES
    3:         15-MINUTES
    4:         20-MINUTES
    5:         30-MINUTES
    6:         60-MINUTES

Select 24-hour storm unit-interval model number..... ===>    "KTYPE"


EFFECTIVE RAINFALL INFORMATION DISPLAY OPTIONS:
    0:        Omit Effective Rainfall information from results
    1:        Include Effective Rainfall information with results
Select display option desired...................... ===>    "KSOIL"



--------------------------------------------------------------------------------
TYPE: EXIT to leave program ; TOP  to go to top of page
                            ; BACK to go back one page
```

```
HYDROGRAPH RESULTS LISTING OPTIONS:

  1: List ENTIRE runoff-hydrograph

  2: List runoff-hydrograph according to USER-SPECIFIED
     time limits

Select desired option.............................. ===>    "SEE BELOW"

Enter model time(HOURS) for BEGINNING of results.... ===>    "TIME1"
:ALLOWABLE VALUES ARE [0.0    ] TO [20.0  ]

Enter model time(HOURS) for END of results.......... ===>    "TIME2"
:ALLOWABLE VALUES ARE [0.0    ] TO [30.0  ]

-----------------------------------------------------------------------------
TYPE: EXIT to leave program ; TOP  to go to top of page
                            ; BACK to go back one page
```

PROGRAM 21

```
C ------------------------------------------------------------------
        SUBROUTINE UNITH
C ------------------------------------------------------------------
C
C  UNIT HYDROGRAPH
C
C
        DIMENSION H(440)
        DIMENSION UH(150),PERCNT(150),R(288),IOFILE(16)
        DIMENSION DYR(48)
        COMMON/NUT/NUT
C
        DATA DYR/0.,20.,40.,60.,80.,100.,125.,150.,100.,89.4,82.3,78.,
     C  74.8,72.4,70.2,68.5,100.,94.8,91.2,88.4,86.3,84.5,82.7,81.3,
     C  100.,96.2,93.8,91.8,90.3,89.,87.8,86.7,100.,97.4,95.8,94.8,
     C  93.9,93.2,92.8,92.3,100.,82.,72.5,66.6,63.,60.8,59.,57.5/
        DATA UH/150*0./
        DATA PERCNT/150*0./
        DATA R/288*0./
        DATA H/440*0./
C
C  FUNCTION
C
        RR(T)=EXP(A*ALOG(T)+B)
C
C..READ DATA INPUT
        READ FREE(5)KTYPE,XL,XLCA,HH,XN,AREA,VSL,KODE1,BASCON,SLP,
     .  R5,R30,R1,R3,R6,R24,SS,KSTORM,KSOIL,PV,PF,PM,PD,NUT,
     .  FX5,FX30,FX1,FX3,FX6,FX24,DAOPT,TIME1,TIME2
        AX=AREA
        XLX=XL
        XLCAX=XLCA
C
        FX=1.
C
        IF(DAOPT.EQ.2)GO TO 183
C
        AREA=AREA/640.
        FX1=.651
        FX3=.78
        FX6=.831
        FX24=.91
        IF(AREA.GE.350.)GOTO 180
C
        DO 100 I=1,13
        IF(AREA.LE.DYR(I))GOTO 110
100     CONTINUE
C
110     I=I-1
        DX=DYR(I+1)-DYR(I)
        FACT=AREA-DYR(I)
        FX1=(DYR( 8+I)-FACT*(DYR( 8+I)-DYR( 9+I))/DX)/100.
        FX3=(DYR(16+I)-FACT*(DYR(16+I)-DYR(17+I))/DX)/100.
        FX6=(DYR(24+I)-FACT*(DYR(24+I)-DYR(25+I))/DX)/100.
        FX24=(DYR(32+I)-FACT*(DYR(32+I)-DYR(33+I))/DX)/100.
        FX30=(DYR(40+I)-FACT*(DYR(40+I)-DYR(41+I))/DX)/100.
        FX5=FX30
180     CONTINUE
        AREA=AREA*640.
```

```
C
C
183     CONTINUE
        WRITE(NUT,5000)
        WRITE(NUT,181)
181     FORMAT(/,14X,'U N I T - H Y D R O G R A P H   A N A L Y S I S',
     .  /)
        WRITE(NUT,5000)
6666    FORMAT(/,10X,'*USER SPECIFIED PRECIPITATION DEPTH-AREA ',
     .  'REDUCTION FACTORS:')
6665    FORMAT(/,10X,'PRECIPITATION DEPTH-AREA REDUCTION FACTORS:')
6664    FORMAT(
     .  12X,'5-MINUTE FACTOR = ',F5.3,/,12X,'30-MINUTE FACTOR = ',
     .   F5.3,/,
     C  12X,'1-HOUR FACTOR = ',F5.3,/,12X,'3-HOUR FACTOR = ',F5.3,/,
     C  12X,'6-HOUR FACTOR = ',F5.3,/,12X,'24-HOUR FACTOR = ',F5.3)
C
C  DETERMINE HYDROGRAPH FACTORS
C
        XL=XL/5280.
        XLCA=XLCA/5280.
      S=HH/XL
      XLAGX=1.2*(XL*XLCA/S**0.5)**0.38
      XLAG=20.0*XN*XLAGX
C
C  DESIGNATE UNIT INTERVALS
C
        IF(KTYPE.EQ.1)UNIT=5.
        IF(KTYPE.EQ.2)UNIT=10.
        IF(KTYPE.EQ.3)UNIT=15.
        IF(KTYPE.EQ.4)UNIT=20.
        IF(KTYPE.EQ.5)UNIT=30.
        IF(KTYPE.EQ.6)UNIT=60.
C
        TIMLAG=100.*UNIT/60./XLAG
        SQMI=AREA*43560./5280./5280.
        BASFLO=BASCON*SQMI
        XK=645.*SQMI*60./UNIT
        XK=XK/100.
C
C  RETURN XL AND XLCA TO UNITS OF FEET
C
        XL=XL*5280.
        XLCA=XLCA*5280.
5000    FORMAT(1X,76('*'))
C
        WRITE(NUT,5000)
        WRITE(NUT,5001)XLX,XLCAX,HH,XN,AX,XLAG,UNIT,TIMLAG,
     C  BASCON,VSL,SLP
5001    FORMAT(10X,'WATERCOURSE LENGTH = ',F11.3,' FEET',/,
     C10X,'LENGTH FROM CONCENTRATION POINT TO CENTROID = ',
     CF11.3,' FEET',/,10X,'ELEVATION VARIATION ALONG ',
     C 'WATERCOURSE = ',F11.3,' FEET',/,10X,
     C 'BASIN FACTOR = ',F5.3,/,
     C 10X,'WATERSHED AREA = ',F11.3,' ACRES',
     C/,10X,'WATERCOURSE "LAG" TIME = ',F8.3,' HOURS',/,
     C10X,'UNIT HYDROGRAPH TIME UNIT = ',F7.3,' MINUTES',/,
     C10X,'UNIT INTERVAL PERCENTAGE OF LAG-TIME = ',F7.3,/,
     C10X,'BASEFLOW = ',F7.3,' CFS/SQUARE-MILE',/,10X,
     C  'MAXIMUM WATERSHED LOSS RATE(INCH/HOUR) = ',F6.3,/,
```

```
     C  10X,'LOW LOSS RATE PERCENTAGE(DECIMAL) = ',F5.3)
        IF(KODE1.EQ.1)WRITE(NUT,8200)
        IF(KODE1.EQ.2)WRITE(NUT,8202)
        IF(KODE1.EQ.3)WRITE(NUT,8204)
        IF(KODE1.EQ.4)WRITE(NUT,8205)
        IF(KODE1.EQ.5)WRITE(NUT,82041)PV,PF,PM,PD
        IF(KODE1.EQ.6)WRITE(NUT,82042)
82041   FORMAT(10X,'VALLEY "S"-CURVE PERCENTAGE(DECIMAL NOTATION) = ',
     C  F6.3,/,10X,'FOOTHILL "S"-CURVE PERCENTAGE(DECIMAL NOTATION) = '
     C  ,F6.3,/,10X,'MOUNTAIN "S"-CURVE PERCENTAGE(DECIMAL NOTATION) = '
     C  ,F6.3,/,10X,'DESERT "S"-CURVE PERCENTAGE(DECIMAL NOTATION) = ',
     C  F6.3,/)
8205    FORMAT(10X,'DESERT S-GRAPH SELECTED',/)
8200    FORMAT(10X,'VALLEY S-GRAPH SELECTED',/)
8202    FORMAT(10X,'FOOTHILL S-GRAPH SELECTED',/)
8204    FORMAT(10X,'MOUNTAIN S-GRAPH SELECTED',/)
82042   FORMAT(10X,'U.S. SOIL CONSERVATION SERVICE S-GRAPH SELECTED',/)
        WRITE(NUT,8209)R5,R30
8209    FORMAT(10X,'SPECIFIED PEAK 5-MINUTES RAINFALL(INCH)= ',F5.2,/,
     C  10X,'SPECIFIED PEAK 30-MINUTES RAINFALL(INCH) = ',F5.2)
        WRITE(NUT,8206)R1,R3,R6,R24,KTYPE
8206    FORMAT(10X,'SPECIFIED PEAK 1-HOUR RAINFALL(INCH) = ',F5.2,/,
     C  10X,'SPECIFIED PEAK 3-HOUR RAINFALL(INCH) = ',F5.2,/,
     C  10X,'SPECIFIED PEAK 6-HOUR RAINFALL(INCH) = ',F5.2,/,
     C  10X,'SPECIFIED PEAK 24-HOUR RAINFALL(INCH)= ',F5.2,//,
     C  10X,'*HYDROGRAPH MODEL #',I1,' SPECIFIED*')
        IF(DAOPT.EQ.1.)WRITE(NUT,6665)
        IF(DAOPT.EQ.2.)WRITE(NUT,6666)
        WRITE(NUT,6664)FX5,FX30,FX1,FX3,FX6,FX24
C
C  USER TIME LIMITS
        IF(TIME2.EQ.31.)GO TO 957
        WRITE(NUT,958)TIME1,TIME2
958     FORMAT(/10X,'RUNOFF HYDROGRAPH LISTING LIMITS:',
     .  /10X,'MODEL TIME(HOURS) FOR BEGINNING OF RESULTS',
     .  ' = ',F5.2,
     .  /10X,'MODEL TIME(HOURS) FOR END OF RESULTS ',
     .  '= ',F5.2)
C
        XR=VSL*UNIT/60.
C
C  DETERMINATION OF UNIT HYDROGRAPH ORDINATES
C
        IF(KODE1.LT.5 .OR. KODE1.EQ.6)GOTO 94
C
C=======LINEAR WEIGHTING OF S-CURVES
        NLAG=0
        DO 98 KLAG=1,4
        PLAG=PV
        IF(KLAG.EQ.2)PLAG=PF
        IF(KLAG.EQ.3)PLAG=PM
        IF(KLAG.EQ.4)PLAG=PD
        IF(KLAG.EQ.0)GOTO 98
        CALL SUBSB(TIMLAG,PERCNT,KLAG,NUMBER)
        IF(NUMBER.GT.NLAG)NLAG=NUMBER
        DO 99 I=1,150
        IF (PERCNT(I).EQ.0.)PERCNT(I)=100.
        H(I)=H(I)+PERCNT(I)*PLAG
 99     PERCNT(I)=0.
 98     CONTINUE
```

```
C
C  UNIT HYDROGRAPH DETERMINATION
C
       NUMBER=NLAG
C
       DO 97 I=1,150
       IF(I.LE.NLAG)PERCNT(I)=H(I)
97     H(I)=0
       GOTO 96
C
94     CALL SUBSB(TIMLAG,PERCNT,KODE1,NUMBER)
C
96    SUM=0.
       IF(NUMBER.GE.150)WRITE(NUT,8207)
8207   FORMAT(3X,'UNIT HYDROGRAPH TERMINATED AFTER 150 UNIT INTERVALS',/)
       IF(NUMBER.GT.150)NUMBER=150
C
       DO 8208 I=1,NUMBER
       UH(I)=(PERCNT(I)-SUM)*XK
       SUM=PERCNT(I)
       IF(UH(I).LT.0.)UH(I)=0.
8208   CONTINUE
C
       WRITE(NUT,5002)
5002   FORMAT(23X,'UNIT HYDROGRAPH DETERMINATION',/)
       WRITE(NUT,5003)
5003   FORMAT(76('-'))
       WRITE(NUT,5004)
5004   FORMAT(5X,'INTERVAL',10X,'"S" GRAPH',10X,'UNIT HYDROGRAPH',/,
     C6X,'NUMBER',10X,'MEAN VALUES',10X,'ORDINATES(CFS)',)
       WRITE(NUT,5003)
       WRITE(NUT,5005)(I,PERCNT(I),UH(I),I=1,NUMBER)
5005   FORMAT(7X,I3,15X,F7.3,13X,F10.3)
C
       WRITE(NUT,5000)
5050   FORMAT(A1)
       IF(KSOIL.EQ.1)WRITE(NUT,5000)
       IF(KSOIL.EQ.1)WRITE(NUT,5020)
       IF(KSOIL.EQ.1)WRITE(NUT,5003)
5020   FORMAT(10X,'UNIT',14X,'UNIT',12X,'UNIT',14X,'EFFECTIVE',/,
     C  9X,'PERIOD',11X,'RAINFALL',7X,'SOIL-LOSS',12X,'RAINFALL',/,
     C  8X,'(NUMBER)',10X,'(INCHES)',8X,'(INCHES)',12X,'(INCHES)')
C
C
C  24-HOUR STORM RAINFALL PATTERN
C
       R5A=R5*FX5
       R30A=R30*FX30
       R1A=R1*FX1
       R3A=R3*FX3
       R6A=R6*FX6
       R24A=R24*FX24
C
C
       A=(ALOG(R30A)-ALOG(R5A))/(ALOG(.5)-ALOG(.0833))
       B=ALOG(R5A)-A*ALOG(.0833)
       R(193)=RR(.0833)
       R(194)=RR(.1667)-RR(.0833)
       R(195)=RR(.25)-RR(.1667)
       R(192)=RR(.3333)-RR(.25)
```

```
      R(196)=RR(.41667)-RR(.3333)
      R(191)=RR(.5)-RR(.41667)
      A=(ALOG(R1A)-ALOG(R30A))/(ALOG(1.)-ALOG(.5))
      B=ALOG(R30A)-A*ALOG(.5)
      R(197)=RR(.5833)-RR(.5)
      R(190)=RR(.6667)-RR(.5833)
      R(198)=RR(.75)-RR(.6667)
      R(189)=RR(.8333)-RR(.75)
      R(188)=RR(.9167)-RR(.8333)
      R(187)=RR(1.)-RR(.9167)
C.......REMAINING PART OF PEAK 3-HOUR STORM
      A=(ALOG(R3A)-ALOG(R1A))/(ALOG(3)-ALOG(1))
      B=ALOG(R1A)-A*ALOG(1)
      RRSAVE=R1A
C
      DO 1001 J=1,12
      XJ=J
      DT=XJ*.1667
      T=1.+DT
      RRNEW=RR(T)
      DR=(RRNEW-RRSAVE)/2.
      R(198+J)=DR
      R(187-J)=DR
1001  RRSAVE=RRNEW
C
C  REMAINING PART OF PEAK 6-HOUR STORM
C
      A=(ALOG(R6A)-ALOG(R3A))/(ALOG(6)-ALOG(3))
      B=ALOG(R3A)-A*ALOG(3)
      RRSAVE=R3A
C
      DO 1010 J=1,18
      XJ=J
      DT=XJ*.1667
      T=3.+DT
      RRNEW=RR(T)
      DR=(RRNEW-RRSAVE)/2.
      R(210+J)=DR
      R(175-J)=DR
1010  RRSAVE=RRNEW
C
C  REMAINING PART OF PEAK 24-HOUR STORM
      A=(ALOG(R24A)-ALOG(R6A))/(ALOG(24)-ALOG(6))
      B=ALOG(R6A)-A*ALOG(6)
      RRSAVE=R6A
C
      DO 1020 J=1,60
      XJ=J
      DT=XJ*.1667
      T=6.+DT
      RRNEW=RR(T)
      DR=(RRNEW-RRSAVE)/2.
      R(228+J)=DR
      R(157-J)=DR
1020  RRSAVE=RRNEW
C
      DO 1030 J=1,96
      XJ=J
      DT=XJ*.08333
      T=16.+DT
```

```
       RRNEW=RR(T)
       DR=RRNEW-RRSAVE
       R(97-J)=DR
1030   RRSAVE=RRNEW
C
C  ADJUST R-ARRAY FOR LARGER UNIT INTERVALS
C
C
C  NI = NUMBER OF STORM HYDROGRAPH INTERVALS
       NI=288
       IF(KTYPE.EQ.1)GOTO 1050
       K=KTYPE
       IF(KTYPE.EQ.5)K=6
       IF(KTYPE.EQ.6)K=12
       NI=288/K
       DO 1040 I=1,NI
       TEMP=0.
       II=(I-1)*K
       DO 1035 J=1,K
1035   TEMP=TEMP+R(II+J)
1040   R(I)=TEMP
1050   CONTINUE
C
C  ADJUST FOR CONSTANT SOIL LOSS
C
       XTOTAL=0.
       TEMPS=UNIT/60.
       XRA=XR
       DO 300 I=1,NI
       XLOSS=R(I)*SLP
       IF(XLOSS.GT.XRA) XLOSS=XRA
       TEMP=R(I)
       R(I)=R(I)-XLOSS
       XTOTAL=XTOTAL+XLOSS
C
       IF(KSOIL.EQ.1)WRITE(NUT,5021)I,TEMP,XLOSS,R(I)
5021   FORMAT(10X,I3,13X,F6.4,11X,F6.4,14X,F6.4)
  300 CONTINUE
C
C DETERMINE STORM RUNOFF HYDROGRAPH
C
       INTERV=NUMBER+NI-1
       IF(INTERV.GT.440)WRITE(NUT,432)
       IF(INTERV.GT.439)INTERV=439
432    FORMAT(3X,'RUNOFF HYDROGRAPH TERMINATED AFTER 440 UNIT',/,
     C 3X,'INTERVALS. SUGGEST USING A LARGER UNIT INTERVAL.')
       DO 600 I=1,INTERV
       M=I
       N=1+M
       DO 500 J=1,M
       K=N-J
       IF(J.GT.NI)GO TO 500
       IF(K.GT.NUMBER) GO TO 500
       H(M)=H(M)+R(J)*UH(K)
  500 CONTINUE
       H(M)=H(M)+BASFLO
  600 CONTINUE
C
C  COMPUTE SUMMED HYDROGRAPH
C
```

```
          SUM=0.0
          XMAX=0.
          DO 700 I=1,INTERV
          IF(H(I).LT.0.)H(I)=0.
          IF(H(I).GT.XMAX)XMAX=H(I)
          SUM=SUM + H(I)
  700   CONTINUE
          SUM=SUM*UNIT*60./43560.
          XTOTAL=XTOTAL/12.*AREA
C
          WRITE(NUT,5003)
          WRITE(NUT,6008)XTOTAL,SUM
6008      FORMAT(5X,'TOTAL SOIL-LOSS VOLUME(ACRE-FEET) = ',F12.4,/,
       C  5X,'TOTAL STORM RUNOFF VOLUME(ACRE-FEET) = ',F12.4)
C
          IF(XMAX.LT.100.)GOTO 8000
          I=XMAX/100.
          II=I+1
          XMAX=II
          XMAX=XMAX*100.
          GOTO 8100
8000      I=XMAX/10.
          II=10*(I+1)
          XMAX=II
8100      CONTINUE
          WRITE(NUT,5003)
C
          CALL OASB(KTYPE,H,INTERV,XMAX,UNIT,SUM,TIME1,TIME2)
C
10000     CONTINUE
C
          RETURN
          END
```

PROGRAM 22

```
C  ----------------------------------------------------------------
      SUBROUTINE SUBSB(TIMLAG,PERCNT,KODE1,NUMBER)
C  ----------------------------------------------------------------
C
C     THIS SUBROUTINE DETERMINES THE PERCNTAGES OF DISCHARGE
C     FACTORS FOR THE VARIOUS ZONE CLASSIFICATIONS.
C
C
C
      REAL MNT
      DIMENSION PERCNT(150),VAL(31,2),FOOT(31,2),MNT(33,2),SCS(33,2)
      DIMENSION DESERT(33,2)
      COMMON/NUT/NUT
C
      DATA VAL/0.,15.0,25.,35.0,50.,65.0,75.,100.,115.,125.,
     C140.0,150.0,165.0,175.0,200.0,225.0,250.,275.,300.,325.,350.,
     C375.,400.,450.,500.,550.,600.,650.,700.,750.,99999.,0.0,2.6,
     C5.0,8.6,15.5,25.0,32.0,50.0,57.9,62.0,66.8,69.5,72.6,74.3,
     C78.0,81.0,83.5,85.7,87.5,89.0,90.5,91.6,92.7,94.3,95.8,
     C96.9,97.8,98.5,99.0,99.5,100. /
      DATA FOOT/0.,15.0,25.,35.0,50.,60.0,75.,85.0,90.0,95.0,
     C100.0,110.0,125.0,140.0,150.0,175.0,200.0,225.0,250.0,
     C275.0,300.0,325.0,350.0,375.0,400.0,450.0,500.0,550.0,
     C600.0,650.0,99999.,0.0,1.9,3.8,6.0,10.3,14.0,21.7,29.0,
     C34.2,43.3,50.0,56.9,63.8,69.0,71.9,77.8,82.4,86.0,89.0,
     C91.4,93.4,95.0,96.2,97.2,97.9,98.5,99.0,99.3,99.7,
     C99.9,100. /
      DATA MNT/0.,15.0,25.,35.0,40.0,50.,65.0,75.0,90.0,
     C100.0,115.0,125.0,140.0,150.0,175.0,200.0,225.0,250.0,
     C275.0,300.0,325.0,350.0,375.0,400.0,450.0,500.0,
     C550.0,600.0,650.0,700.0,750.0,800.0,99999.0,0.0,
     C3.3,6.7,10.6,13.4,21.0,33.0,39.3,46.3,50.0,54.2,
     C56.7,59.8,61.8,65.8,69.2,72.2,74.8,77.0,79.0,80.7,
     C82.2,83.5,84.8,86.9,88.9,90.5,92.0,93.3,94.5,95.5,
     C96.4,100. /
      DATA DESERT/0.,12.5,25.0,37.5,50.0,62.5,75.0,87.5,100.,
     C112.5,125.,137.5,150.,162.5,175.,187.5,200.,225.,250.,275.,
     C300.,325.,350.,375.,400.,450.,500.,550.,600.,700.,800.,1000.,
     C9999.,0.,1.1,3.2,6.3,10.5,18.5,31.3,42.0,50.0,56.5,61.3,65.2,
     C68.5,71.5,74.0,76.2,78.3,81.6,84.3,86.7,88.7,90.2,91.6,92.8,
     C93.9,95.6,96.9,97.8,98.3,99.5,99.9,99.99,100./
      DATA SCS/0.,8.6,17.2,25.9,34.5,43.1,51.7,60.3,69.,77.6,
     C 86.2,94.8,103.4,112.1,120.7,129.3,137.9,146.5,155.2,
     C 163.8,172.4,189.6,206.9,224.1,241.4,258.6,275.8,293.1,
     C 310.3,327.6,344.8,387.9,999.,0.,.1,.6,1.2,3.5,6.5,10.7,
     C 16.3,22.8,30.,37.5,45.,52.2,58.9,65.,70.,75.1,79.,82.2,
     C 84.9,87.1,90.8,93.4,95.3,96.7,97.7,98.4,98.9,99.3,99.5,99.7,
     C 99.9,100./
C
C     KODE1=1: VALLEY
C     KODE1=2: FOOTHILL
C     KODE1=3: MOUNTAIN
C     KODE1=4: DESERT
C     KODE1=5: NOT USED DIRECTLY.  LINEAR INTERPOLATION OF 1-4
C     KODE1=6: SCS
C     TIMLAG=LAG
C     TIME=INCREMENTS OF LAG
C     NUMBER=INDEX OF PERCNT VECTOR
C
```

```
      ANEW=0.
      AOLD=0.
      K=1
      NUMBER=1
      TIME=TIMLAG
10    K=K+1
      IF(NUMBER.GE.151)GO TO 1000
      N=K-1
C
      GO TO(100,200,300,350,355,355)KODE1
C
100   CONTINUE
C
C     INTEGRATE "S" GRAPH IN ORDER TO DETERMINE UH(I)
C
      TEMP=0.5*(VAL(K,2)+VAL(N,2))*
     C (VAL(K,1)-VAL(N,1))
      ANEW=ANEW+TEMP
      IF(TIME.GT.VAL(K,1))GO TO 10
      Y=VAL(K,2)
      X=VAL(N,2)
      DEL=(TIME-VAL(N,1))/(VAL(K,1)-VAL(N,1))
      B=VAL(K,1)-VAL(N,1)
      GO TO 400
200   CONTINUE
      TEMP=0.5*(FOOT(K,2)+FOOT(N,2))*
     C (FOOT(K,1)-FOOT(N,1))
      ANEW=ANEW+TEMP
      IF(TIME.GT.FOOT(K,1))GO TO 10
C
      Y=FOOT(K,2)
      X=FOOT(N,2)
      DEL=(TIME-FOOT(N,1))/(FOOT(K,1)-FOOT(N,1))
      B=FOOT(K,1)-FOOT(N,1)
      GO TO 400
300   CONTINUE
      TEMP=0.5*(MNT(K,2)+MNT(N,2))*
     C (MNT(K,1)-MNT(N,1))
      ANEW=ANEW+TEMP
      IF(TIME.GT.MNT(K,1))GO TO 10
      Y=MNT(K,2)
      X=MNT(N,2)
      DEL=(TIME-MNT(N,1))/(MNT(K,1)-MNT(N,1))
      B=MNT(K,1)-MNT(N,1)
C
      GO TO 400
350   CONTINUE
      TEMP=0.5*(DESERT(K,2)+DESERT(N,2))*
     C (DESERT(K,1)-DESERT(N,1))
      ANEW=ANEW+TEMP
      IF(TIME.GT.DESERT(K,1))GO TO 10
      Y=DESERT(K,2)
      X=DESERT(N,2)
      DEL=(TIME-DESERT(N,1))/(DESERT(K,1)-DESERT(N,1))
      B=DESERT(K,1)-DESERT(N,1)
      GO TO 400
C
355   CONTINUE
C
C SCS METHOD
```

```
C
      TEMP=0.5*(SCS(K,2)+SCS(N,2))*(SCS(K,1)-SCS(N,1))
      ANEW=ANEW+TEMP
      IF(TIME.GT.SCS(K,1)) GO TO 10
      Y=SCS(K,2)
      X=SCS(N,2)
      DEL=(TIME-SCS(N,1))/(SCS(K,1)-SCS(N,1))
      B=SCS(K,1)-SCS(N,1)
C
400   CONTINUE
      DEL=DEL*(Y-X)
      XX=X+DEL
C
C     ADJUST INTEGRATION FOR INTERPOLATION
C
      DELA=0.5*(Y+XX)*(1.-DEL/(Y-X))*B
      ANEW=ANEW-DELA
      PERCNT(NUMBER)=(ANEW-AOLD)/TIMLAG
      NUMBER=NUMBER+1
      AOLD=ANEW
      ANEW=ANEW-.5*(X+XX)*DEL/(Y-X)*B
      TIME=TIME+TIMLAG
      K=K-1
      IF(NUMBER.EQ.2)GO TO 10
      DELX=PERCNT(NUMBER-1)-PERCNT(NUMBER-2)
      IF(DELX.LE..51)GO TO 1000
      GO TO 10
1000  CONTINUE
      NUMBER=NUMBER-1
      IF(NUMBER.GE.150)WRITE(NUT,1001)NUMBER
1001  FORMAT(7X,'UNIT-HYDROGRAPH TERMINATED AFTER',I4,' INTERVALS')
C
C  FINISH-OFF HYDROGRAPH
C
      IF(NUMBER.GE.150)GOTO 1250
      NNUM=150-NUMBER
      XNUM=NNUM
      REM=100.-PERCNT(NUMBER)
      REM1=REM/XNUM
      DELX=.5
      IF(REM1.LT.DELX)GOTO 1150
      DO 1140 K=1,NNUM
1140  PERCNT(NUMBER+K)=PERCNT(NUMBER+K-1)+REM1
      GOTO 1200
1150  XNUM=REM/DELX+1.
      NNUM=XNUM
      DO 1160 K=1,NNUM
1160  PERCNT(NUMBER+K-1)=PERCNT(NUMBER+K-2)+DELX
      NNUM=NNUM-1
1200  NUMBER=NUMBER+NNUM
      IF(PERCNT(NUMBER).GE.100.)PERCNT(NUMBER)=100.
      IF(NUMBER.GE.100)GOTO 1250
      IF(PERCNT(NUMBER).GE.100.)GOTO 1250
      NUMBER=NUMBER+1
      PERCNT(NUMBER)=100.
1250  CONTINUE
C
      RETURN
      END
```

PROGRAM 23

```
C  -----------------------------------------------------------------------
        SUBROUTINE OASB(KTYPE,H,INTERV,XMAX,UNIT,SUM,TIME1,TIME2)
C  -----------------------------------------------------------------------
C
        COMMON/NUT/NUT
        DIMENSION H(440)
        INTEGER LETM,BLANK,DOT,CROSS,DASH,LINE(41)
        DATA LETM,BLANK,DOT,CROSS,DASH,LINE/'V',' ','.','Q','I',41*' '/
C
C
5050    FORMAT(1X,A1)
        WRITE(NUT,101)
        WRITE(NUT,10)
10      FORMAT(/,25X,'2 4 - H O U R    S T O R M',/,
     .  21X,'R U N O F F    H Y D R O G R A P H',/)
100     CONTINUE
        WRITE(NUT,101)
101     FORMAT(76('*'))
102     FORMAT(76('-'))
        STEP=5.*60./43560.
        WRITE(NUT,103)
103     FORMAT(18X,'HYDROGRAPH IN FIVE-MINUTE INTERVALS(CFS)',/)
        WRITE(NUT,102)
        F=40./XMAX
        FMASS=40./SUM
        T0=0.0
        T1=XMAX/4.
        T2=2.*T1
        T3=3.*T1
        WRITE(NUT,105)T0,T1,T2,T3,XMAX
105     FORMAT(2X,'TIME(HRS) VOLUME(AF)    Q(CFS) ',F2.0,1X,4F10.1)
        WRITE(NUT,102)
        XMASS=0.
C
        GO TO(141,142,143,146,147,148),KTYPE
141     KNUM=1
        GOTO 144
142     KNUM=2
        GOTO 144
143     KNUM=3
        GO TO 144
146     KNUM=4
        GO TO 144
147     KNUM=6
        GO TO 144
148     KNUM=12
144     TIME=0.
C
C   OUTPUT GRAPH LOOP
C
        DO 200 I=1,INTERV
        TEST=H(I)*F
        XMASS=XMASS+H(I)*STEP
        LINE(1)=DOT
        LINE(11)=DOT
        LINE(21)=DOT
        LINE(31)=DOT
        LINE(41)=DOT
        J=TEST+1
```

```
      JMASS=XMASS*FMASS+1
      LINE(JMASS)=LETM
      LINE(J)=CROSS
      IF(KNUM.EQ.1)GOTO 400
C
320   DO 350 K=1,KNUM
      TIME=TIME+.083333
      IF(K.EQ.1)GOTO 349
      XMASS=XMASS+H(I)*STEP
      JMASS=XMASS*FMASS+1
      LINE(JMASS)=LETM
      LINE(J)=CROSS
349   IF(TIME.GE.TIME1.AND.TIME.LE.TIME2)WRITE(NUT,210)TIME,XMASS,
     .H(I),LINE
350   LINE(JMASS)=BLANK
      GOTO 215
C
400   CONTINUE
      TIME=TIME+.083333
      IF(TIME.GE.TIME1.AND.TIME.LE.TIME2)WRITE(NUT,210)TIME,XMASS,
     .H(I),LINE
210   FORMAT(2X,F7.3,F12.4,1X,F9.2,2X,41A1)
215   LINE(J)=BLANK
      LINE(JMASS)=BLANK
200   CONTINUE
      WRITE(NUT,101)
C
      RETURN
      END
```

CHAPTER 6.

INTERACTIVE DESIGN

The use of computers to aid in water resources related analysis, synthesis, and design has increased significantly during the last decade. A main motivation for computer use is that water resources related studies often require (1) an iterative calculation analog such as that used in the calculation of hydraulic section information, (2) solution of a convolution type integral such as is used in unit-hydrograph hydrology studies, or (3) the solution of a simultaneous system of equations such as is employed in water distribution network analysis. Because each of these three general classifications of problems essentially involve a repetitive series of calculations, a computer code can be prepared which will offer the engineer an extremely cost effective tool.

Another motivation for the use of computers in water resources related studies is the development and widespread use of digital microcomputers. For many classes of problems, the microcomputer offers the speed and capability to the single user as does a minicomputer in a multiple user system. Consequently, programming techniques which were once limited to the minicomputer or the mainframe class of computers is now available at low cost by means of a microcomputer system.

Such programming techniques include 'humanized' computer interaction and detailed, easy to read computer results which are explicit, fit the requirements of a reviewing agency, and yet are understandable to the first-time reviewer of the product.

By making the program humanized, the learning curve is essentially minimized in that all program information is prompted and scanned for acceptability and can be rejected if data is not within program specified limits. In general, the user's manuals associated with a wide variety of batch programs are eliminated because the humanized program guides the user through every possible logic path, providing the user with various checks and controls in order to further reduce parameter selection errors and an unreasonable choice of design options. The development of the humanized software may be defined as a uniform communication/presentation (C/P) to the program user. There are several C/P requirements for computer software which are described as follows:

1. For a sequence of data entry prompts, the page number and calculation model description appears at the top of the page.

2. All words are written in their entirety, with abbreviations avoided whenever possible.

3. All units should be given for the information requested.

4. Allowable values are identified which limit the data entries to reasonable quantities.

5. Should the agency manuals suggest criteria for data entry, the recommendations are included on the display.

6. Any program operation commands should be consistently and uniformly displayed so that the user can operate the interaction or special data editing features without confusion.

7. A 'failsafe' line appears on the CRT screen for each page. This line is located near the bottom of the screen and appears in the same position on each page. Below the failsafe line occurs all program operating instructions. These instructions can be typed by the program user at anytime and will cause the program to respond instantly.

8. All computer-dependent requirements (such as data file manipulation) for data management should be interior of the software so the program user need not be knowledgeable of computer operations to use the program.

9. The C/P should be uniform between programs. By requiring uniformity of software interaction, individual programmer personality traits can be avoided. The result is a library of software wherein each program operates, interacts, and responds identically. Consequently, the user learning curve is minimized.

By using a program which provides an optimum product, the usual design review procedure is minimized which reduces total cost to both the design engineer and the design review team. Additionally, the computer product should be the actual fully prepared report which is to be submitted, containing the usual introductory pages, and the study results should be produced in the reviewing agency's required computer-printed forms or plotted graphs. The computer program then provides an actual engineering product, minimizing the need for secretarial and graphic efforts.

History and Concepts of Computer-Human Communications

In the past, the typical manner of communication with a computer was by means of punched computer cards. The cards were returned to the user along with a printed output which sometimes only indicated how well the data deck was assembled. The process was repeated until the desired result was obtained. This type of interaction with a computer is extremelv inefficient and promotes only one direction of communication between the computer and user.

A marked improvement in computer-human communications was the utilization of the cathode ray terminal (CRT) for the data entry via the keyboard and program access. This gave the user a 'hands on' approach with the computer and promoted the development of a conversational relationship. A user could now submit his own jobs, view the output on the terminal screen instead of waiting for a hard copy print, correct the errors if any, and re-submit the job. This type of computer-user interaction greatly increased the efficiency of the user and the cost effectiveness of the digital computer.

The interactive display presented to the user on the terminal by the application program was usually of a type called scrolling. Scrolling is presenting a line of characters or text on the bottom of the terminal screen. This line moves upwards continually as new lines are presented. This type of presentation differs from natural reading techniques such as reading a book in that the material moves upward instead of the eyes moving downward across a steady display.

When using the scroll technique in data entry situations, the user is not aware of the next input requirements until it actually appears on the bottom of the screen. If errors occur while entering data, messages are displayed and scroll up while repeat prompts for user input are again requested by the program. Multiple occurrences of errors usually result in a screen full of scrolling error messages which often add to the confusion. Normally, scrolling interaction never allows a user to change a data entry once it is accepted by the program unless the user restarts the program (thus losing all previously entered data) or edits the data if the capability exists within the program.

In contrast, humanized form-fill out display interaction closely simulates natural reading or viewing characteristics. All textual information or data entry requests are assembled in logically related groups to fit comfortably on 'pages'. These pages are presented to the user by clearing the CRT screen of all previous information before displaying the current 'page'. The text information is displayed starting at the top of the screen and proceeds downward until the bottom limit of the screen is reached. The user observes a stable screen of related information. In the case of data entry, related requests for user input are displayed simultaneously, enabling the user to foresee subsequent data entry requests. The cursor, which can be described as a small rectangular beam of light or pointer that emits characters on the screen, moves down the screen after each input request is satisfied.

This cursor movement is the key to powerful screen interaction. A typical terminal screen can be visualized as a matrix containing 24 lines by 80 columns or 1920 elements (see Figure 6.1). One may individually address these elements by programmed movement of the cursor to the selected element. Textual sequences can be displayed anywhere on the screen at any time while retaining or erasing previous information. This allows warning signals or error messages to appear next to the data in question and disappear after the error has been corrected, thus leaving the original page of displayed information intact (see Figures 6.2, 6.3 and 6.4). Another powerful use of cursor movement is in the case of data entry on a page affecting the permissible values of subsequent data entries on the same page. The allowable values displayed under a subject data entry prompt can be cursor addressed, changed to new values, and the cursor returned to the previous selection instantly. This method employs full conversational awareness by the computer system at all times, and differs from other techniques that 'disconnect' the system with the terminal until a full page of user data is filled out. This later method is called buffered form-fill out because the system is not aware of anything that is entered on the screen until the entire page is returned for processing. This method is better than scroll but not nearly as effective as the C/P interactive method in discussion.

Since the viewing displays are constructed in 'pages', the user can manipulate the pages by a set of understandable interaction commands listed in the following:

1. TOP: request to clear the screen, re-display the page and return the cursor to the first input request on the current page. This permits modification of data on the page.

2. BACK: returns to the previous page for corrections or changes.

3. MAIN: terminates the program function in progress, manipulates computer files as needed, and returns to the main program menu of available processes.

4. EXIT: terminates the program function in progress, closes all computer files as needed, and exits the program.

User directed page movements, coupled with the dynamic display qualitics attained through programmed addressable cursor movements, provide a powerful and flexible interactive environment for experienced as well as first time computer program users.

Considerations for Interactive Computer Software

An important consideration for selecting an interactive design method is compatibility among various computer hardware.

```
12345678901234567890123456789012345678901234567890123456789012345678901234567890
2        1         2         3         4         5         6         7         8
3
4
5
6
7
8
9
10
11
12                        1920 CELLS IN SCREEN MATRIX
13
14
15
16
17
18
19
20
21
22
23
24
```

Fig. 6.1.

```
    ---DATA ENTRY FOR IRREGULAR CHANNEL WATER SURFACE PROFILE---PAGE 1

    In order to determine a water surface profile in a sub-
    critical channel, the channel needs to be defined by
    cross-sections.  These sections are entered as one tra-
    vels upstream.
    Enter TOTAL NUMBER of cross-sections................ ===>█
    :ALLOWABLE VALUES ARE [2] TO [20    ]

    Enter TOTAL irregular channel FLOW(CFS)............. ===>
    :ALLOWABLE VALUES ARE [1     ] TO [999999]

    Enter allowable ERROR(FEET) in water surface
    determination....................................... ===>
    :ALLOWABLE VALUES ARE [.01   ] TO [.50   ]





--------------------------------------------------------------------------------
    TYPE: EXIT to leave program ; TOP  to go to top of page
```

Fig. 6.2.

```
---DATA ENTRY FOR IRREGULAR CHANNEL WATER SURFACE PROFILE---PAGE 1

In order to determine a water surface profile in a sub-
critical channel, the channel needs to be defined by
cross-sections.  These sections are entered as one tra-
vels upstream.
Enter TOTAL NUMBER of cross-sections................ ===>█        *ERROR*
:ALLOWABLE VALUES ARE [2] TO [20    ]                    ^RE-ENTER

Enter TOTAL irregular channel FLOW(CFS)............. ===>
:ALLOWABLE VALUES ARE [1     ] TO [999999]

Enter allowable ERROR(FEET) in water surface
determination....................................... ===>
:ALLOWABLE VALUES ARE [.01    ] TO [.50    ]




*****VALUE ENTERED IS NOT WITHIN ACCEPTABLE RANGE.
--------------------------------------------------------------------------------
TYPE: EXIT to leave program ; TOP  to go to top of page
```

Fig. 6.3.

```
---DATA ENTRY FOR IRREGULAR CHANNEL WATER SURFACE PROFILE---PAGE 1

In order to determine a water surface profile in a sub-
critical channel, the channel needs to be defined by
cross-sections.  These sections are entered as one tra-
vels upstream.
Enter TOTAL NUMBER of cross-sections................ ===>5
:ALLOWABLE VALUES ARE [2] TO [20    ]

Enter TOTAL irregular channel FLOW(CFS)............. ===>█
:ALLOWABLE VALUES ARE [1     ] TO [999999]

Enter allowable ERROR(FEET) in water surface
determination....................................... ===>
:ALLOWABLE VALUES ARE [.01    ] TO [.50    ]




--------------------------------------------------------------------------------
TYPE: EXIT to leave program ; TOP  to go to top of page
```

Fig. 6.4.

Unfortunately, wide differences still exist between manufacturers of peripheral devices such as terminals and computer resident system programs called operating systems. Developing an interactive design methodology that is dependent upon a particular type of hardware or the operating system of a certain computer generally promotes the eventual demise of the approach. The constant change of operating systems due to computer vendor upgrades may render the programming required to accomplish such a design incompatible with the revised system.

More likely to occur is the typical change or upgrades of CRT devices or terminals which may not accommodate some of the interactive features programmed for the previous device. The hardware or operating system dependent functions can be justified in the case where the application software and hardware are bundled together to form a functional package. These types of packages normally deal with graphics applications sych as CAD/CAM systems or elaborate word processing systems. The special purpose computers are mostly self-contained systems requiring certain hardware and operating system configurations.

An interactive design method has to merge comfortably with the major application systems that are already on the machine such as engineering, accounting information retrieval systems, and word processing. In order to accomplish this task and maintain compatability among a wide selection of terminals and operating systems, a certain subset or core group of interactive functions are developed which will perform all the major tasks of a humanized form-filled interactive method. The basic functions that are compatible with over 95% of the terminal hardware and computer systems available today are absolute cursor addressing, clearing the screen, and ringing the bell.

Another requirement is the ability of the computer system to send out what is called control characters which excite these functions. Using these basic functions as building blocks, an entire sophisticated 'humanized' interactive approach can be accomplished while still retaining compatibility across vendor lines. Changes in terminal control character sequences for various terminals are accomplished by an easily accessible hardware table contained within the interactive application driver routines or by table files. These tables map the function to the device. There are many other functions available in terminals such as highlighting or dimming or certain groups of text, flashing of warning messages, and split screening for multiple tasks. These are not normally available in all terminals and are usually reserved for the higher priced models. In addition, the particular code sequences to start and stop these functions are widely different. These functions are just extensions of the basic three required to develop such a user friendly system. A well designed interactive system using the basic three functions of cursor addressing, clear screen, and bell satisfies fully the criteria required to produce a truly 'humanized' interactive methodology.

Screen Layout

As discussed in a previous section, the CRT screen is divided into a cell matrix of 80 columns by 24 rows. The screen layout has been designed to accomodate most data entry possibilities. The basic skeleton screen is set up as follows: (see Fig. 6.5)

Columns on CRT Screen

1 - 4	Blank
5 - 56	All text prompts and allowable values start in column 5 and end in column 56. Additional lines may be used to continue text
57	Blank
58 - 61	Reserved for the '===>' designation which pulls the viewers eye to where the data should be entered
62 - 72	Used for the actual data input
73 - 79	Reserved for the '*ERROR*' designation when input errors are detected.

Rows on CRT screen

21	Reserved for explanatory error messages or engineering warning messages
22	Contains the 'failsafe' line of dashes or underscores
23 - 24	Contains the failsafe program instructions that vary from page to page

Using this uniform layout of screen design will promote adaptability for users of other systems using the same design. Once a user knows how to manipulate one program, the others are basically the same except for the actual data entry prompts dependent on that calculation model.

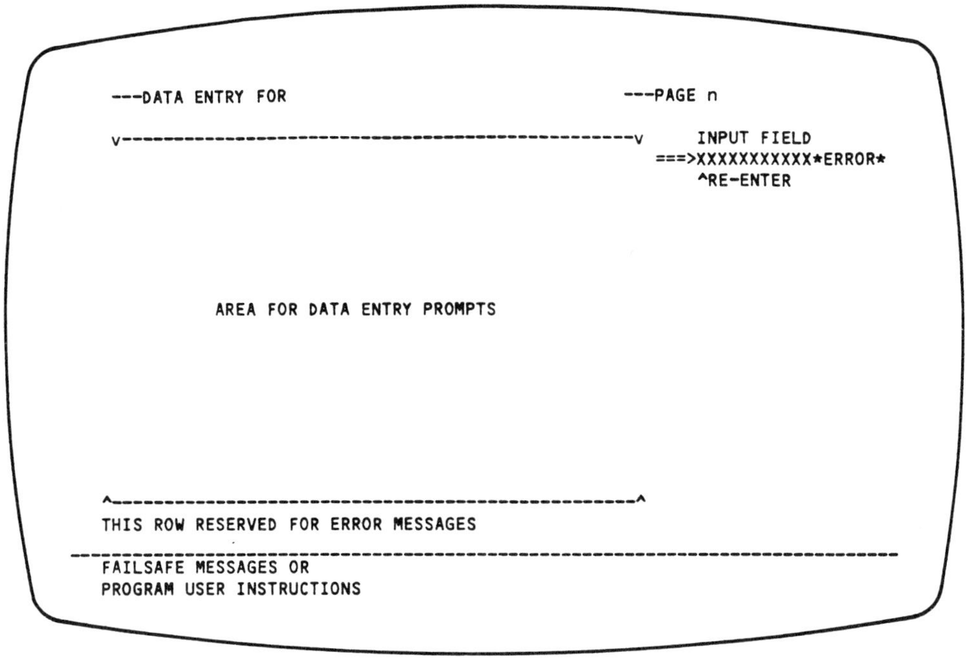
---DATA ENTRY FOR ---PAGE n
v---v
INPUT FIELD
===>XXXXXXXXXXX*ERROR*
^RE-ENTER
AREA FOR DATA ENTRY PROMPTS
^---^
THIS ROW RESERVED FOR ERROR MESSAGES
FAILSAFE MESSAGES OR
PROGRAM USER INSTRUCTIONS

Fig. 6.5.

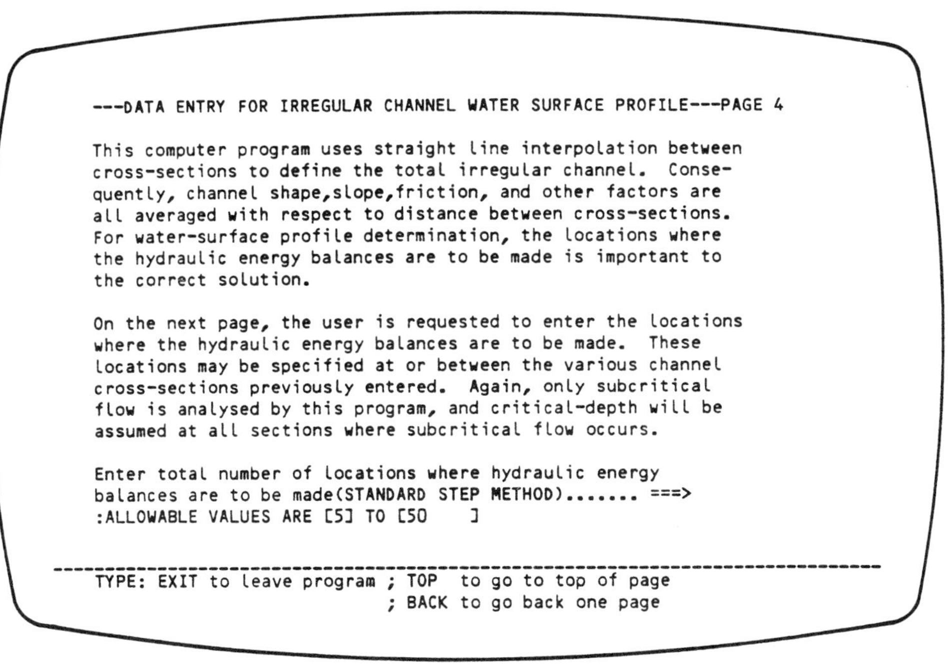
---DATA ENTRY FOR IRREGULAR CHANNEL WATER SURFACE PROFILE---PAGE 4
This computer program uses straight line interpolation between
cross-sections to define the total irregular channel. Conse-
quently, channel shape,slope,friction, and other factors are
all averaged with respect to distance between cross-sections.
For water-surface profile determination, the locations where
the hydraulic energy balances are to be made is important to
the correct solution.
On the next page, the user is requested to enter the locations
where the hydraulic energy balances are to be made. These
locations may be specified at or between the various channel
cross-sections previously entered. Again, only subcritical
flow is analysed by this program, and critical-depth will be
assumed at all sections where subcritical flow occurs.
Enter total number of locations where hydraulic energy
balances are to be made(STANDARD STEP METHOD)....... ===>
:ALLOWABLE VALUES ARE [5] TO [50]
TYPE: EXIT to leave program ; TOP to go to top of page
; BACK to go back one page

Fig. 6.6.

Flow of User Data

The user-entered data flows through a library of form-fill out routines where it is checked for validity, allowable ranges, and real or integer values limits before it can be accepted by the calling calculation subroutine. Detailed descriptions of the subroutines and arguments will be discussed in a subsequent section, but the overall logic flow of user entered data through the subroutines is as follows. All input from the user is read from subroutine GETVAL. This routine is the central or main controlling routine for all input processing. Within GETVAL, various checks are made to validate the data. One set of checks determines if TOP, BACK, EXIT, or MAIN has been typed by the user. In addition, NUMCK is called to determine if a number entered by the user is valid. For example, two decimal points or a '+' or a '-' within the same value would trigger an error message. The error messages are presented by a call to ERROR within GETVAL to display the appropriate message. A call to CLEAN erases the messages on the screen after the user enters correct data. Argument values returned by GETVAL to the calling calculation subroutine are processed to allow continuation of data entry from the user or to jump to logic that will return to a previous data entry page, exit the program, erase the current page and request resubmittal, or go to the main menu.

Using the library of CRT screen handling routines reduces considerably the amount of computer code required for input processing. Since the arguments feeding GETVAL setup permissible ranges, and other parameters unique to a certain input item, the only READ statement for input (for all input items) is contained within GETVAL. All the logic that processes errors and displays messages is contained within ERROR and CLEAN. The area or column-row address of the CRT screen where the current input item resides is passed through all the form-fill out routines in order that error messages and re-enter prompts can be addressed to the CRT correctly. In this general fashion, all input following the basic skeleton frame for the CRT 'pages' can be processed very efficiently.

Screen Design Method

The design of CRT interaction pages that will be presented to the program user for data entry requires a thorough design process in itself to be effective. An approach which has been used successfully in implementing production quality software is described in this section. The first step in the screen design process is to define all input entry prompts for a given function or calculation model. This is absolutely necessary in order that the screen designer may group related data entry prompts in a logical fashion. This global definition phase of prompts also helps to eliminate redundant data entries.

The wording of the prompt for single data items or value should start with 'Enter' followed by the information request text which is then followed by the 'units' designation. Data entry prompts which require a choice among several listed options demands a different strategy. The general title or description of the options is written first. The options are indented and listed below the title. Finally, the prompt starting with 'enter' or 'select' is written which describes the value being requested.

Any descriptive text that will precede prompts excluding the actual prompt must be defined for all input requests. The descriptive text is extremely helpful to first time or occasional users. The text should be very concise and offer a description of what is to follow in the prompt (see Fig. 6.6). After additional explanatory text is defined, all prompts must be inspected and assigned allowable values. This range of values should be set to restrict each prompt to a typical or normal range of values that the user may enter. For option selection prompts which provide choices, only those stated values are permitted. A value outside this range represents an incorrect data response and will be handled by an error message and corresponding re-enter request for data. After all admissible values are defined, any suggested values that may help the user should be defined for each prompt. These values may be typical values given a certain criteria or possible a county-dependent suggested value for a certain input item.

After all the data entry prompts have been defined along with any descriptive text to preclude the prompt, allowable values, and suggested values, the screen designer may now group related items together to fit comfortably on CRT 'pages' (see Fig. 6.7 for a sample CRT page which illustrates the suggested spacing between prompts, descriptive text and allowable values). Figure 6.8 illustrates the computer code to produce the CRT display screen presented in Fig. 6.7. This code implementation may be used as a guide for development of a complete system of user-friendly or humanized form-fill out data entry displays.

Subroutine Descriptions and Listings

SUBROUTINE CRINIT

Abstract

CRINIT initializes the terminal control tables for functions of clear screen, cursor addressing, and bell.

Description

The routine contains all codes necessary to set up the screen functions. After the control codes have been determined for the

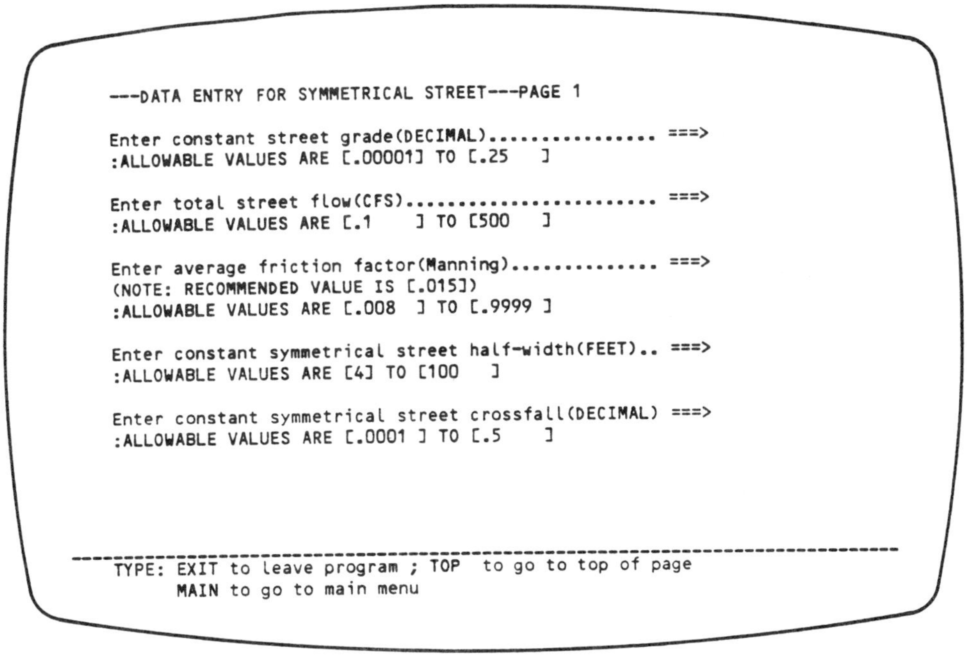
---DATA ENTRY FOR SYMMETRICAL STREET---PAGE 1
Enter constant street grade(DECIMAL)................ ===>
:ALLOWABLE VALUES ARE [.00001] TO [.25]
Enter total street flow(CFS)........................ ===>
:ALLOWABLE VALUES ARE [.1] TO [500]
Enter average friction factor(Manning)............. ===>
(NOTE: RECOMMENDED VALUE IS [.015])
:ALLOWABLE VALUES ARE [.008] TO [.9999]
Enter constant symmetrical street half-width(FEET).. ===>
:ALLOWABLE VALUES ARE [4] TO [100]
Enter constant symmetrical street crossfall(DECIMAL) ===>
:ALLOWABLE VALUES ARE [.0001] TO [.5]
TYPE: EXIT to leave program ; TOP to go to top of page
MAIN to go to main menu

Fig. 6.7.

```
          CALL CLRSCR
          CALL CURSOR(5,1)
          WRITE(WT,706)PG1
          CALL CURSOR(5,3)
          WRITE(WT,707)ARW
          CALL ALLOW(3,4,'.00001     ','.25        ')
          CALL CURSOR(5,6)
          WRITE(WT,708)CFS,ARW
          CALL ALLOW(3,7,'.1         ','500        ')
          CALL CURSOR(5,9)
          WRITE(WT,709)ARW
          CALL ALLOW(3,11,'.008       ','.9999      ')
          CALL CURSOR(5,13)
          WRITE(WT,710)CS,FT,ARW
          CALL ALLOW(1,14,'4          ','100        ')
          CALL CURSOR(5,16)
          WRITE(WT,711)CS,ARW
          CALL ALLOW(3,17,'.0001      ','.5         ')
          CALL INFO(2)
          CALL CURSOR(11,24)
          WRITE(WT,680)
C
C..S0
          CALL GETVAL(62,3,0,.00001,.25,VAL,NCOND,4,0)
          IF(NCOND.NE.0)GO TO(9999,5,5,10000),NCOND
          S0=VAL
C
C..QQ
          CALL GETVAL(62,6,0,.1,500.,VAL,NCOND,4,0)
          IF(NCOND.NE.0)GO TO(9999,5,5,10000),NCOND
          QQ=VAL
C
C..XN
          CALL GETVAL(62,9,0,.008,.9999,VAL,NCOND,4,0)
          IF(NCOND.NE.0)GO TO(9999,5,5,10000),NCOND
          XN=VAL
C
C..HWIDTH
          CALL GETVAL(62,13,0,4.,100.,VAL,NCOND,4,0)
          IF(NCOND.NE.0)GO TO(9999,5,5,10000),NCOND
          HWIDTH=VAL
C
C..XFALL
          CALL GETVAL(62,16,0,.0001,.5,VAL,NCOND,4,0)
          IF(NCOND.NE.0)GO TO(9999,5,5,10000),NCOND
          XFALL=VAL
C
C  FORMATS
C
706       FORMAT('---DATA ENTRY FOR SYMMETRICAL STREET',
     .    '---',A6,Z)
707       FORMAT('Enter constant street grade(DECIMAL)',16('.'),A5,Z)
708       FORMAT('Enter total street flow',A5,24('.'),A5,Z)
709       FORMAT('Enter average friction factor(Manning)',14('.'),
     .    A5,/4X,'(NOTE: RECOMMENDED VALUE IS [.015])',Z)
710       FORMAT('Enter constant symmetrical street half-width',A6,'..',
     .    A5,Z)
711       FORMAT('Enter constant symmetrical street crossfall(DECIMAL)',
     .    A5,Z)
680       FORMAT('MAIN to go to main menu',Z)
```

Fig. 6.8.

terminal selected, these values are then loaded into the IP and ICLEAR arrays. The IP array contains the ASCII value required for positioning the cursor from 1 thru 80 on the CRT screen. ICLEAR array contains the ASCII codes to clear the screen. IESC is loaded with the lead in character for the cursor addressing sequence. IBELL contains the code to exite the bell function on the terminal. The terminal codes set up in the listing are compatible with Lear Sigler, Televideo, and SOROC terminals. The ASCII values of the terminal functions are stored in common block CURS for availability throughout the library of screen subroutines. Routine is called once to set up the codes.

Arguments

None.

SUBROUTINE CURSOR

Abstract

Positions the cursor anywhere on the CRT screen.

Description

Receives arguments of column, row from calling subroutine to position the cursor before a write is performed on the screen. The column, row argument is used as indexes to the IP array table to select the appropriate ASCII codes. The write statement submits the ASCII string of characters to the screen for cursor positioning. The Z (computer dependent) in the format statement holds the cursor after postioning for subsequent writes or prompts. Some systems will not need the Z type of specification in the format statement.

Arguments

IX	(IN)	Column number (1-80)
IY	(IN)	Row number (1-24)

SUBROUTINE GETVAL

Abstract

Central data input capture routine which gets and returns a legal value from user input.

Description

Input is received in an 11-character buffer (BUF) in Alpha format. The buffer is first checked for any legal function such as TOP, BACK, or EXIT. Depending on MODE, certain functions will not be legal at certain times. NUMCK is called to transform the alpha characters into a legal real value. If an error occurs, ERROR is called to display a message. The logic remains in GETVAL until the error is corrected, after which CLEAN is called to erase the previous error messages.

Arguments

IX	(IN)	Column number to read on CRT screen
IY	(IN)	Row number to read on CRT screen
INT	(IN)	0 = Read data input anticipated 1 = Integer data input anticipated
FMIN	(IN)	Minimum allowable value-passed to NUMCK
FMAY	(IN)	Maximum allowable value-passed to NUMCK
VAL	(OUT)	Legal, checked real value returned to calling routine
NCOND	(OUT)	1 = EXIT input by user 2 = TOP input by user 3 = BACK input by user 4 = MAIN input by user Values returned to calling routine for appropriate action
MODE	(IN)	1 = allow EXIT 2 = allow EXIT, TOP 3 = allow EXIT, TOP, BACK 4 = allow EXIT, TOP, MAIN
NULL	(IN)	0 = do not allow blanks as input 1 = allow blanks as input, but set VAL = 0 at this occurrence

SUBROUTINE NUMCK

Abstract

Extensively checks a data input value (numeric) for legal syntax.

Description

The user data input is received in alpha format in buffer KFLD. KFLD is then parced character by character to determine a legal numeric form. If the form is legal, KFLD is transformed by ENCODE and DECODE operations to a REAL value. If an integer is expected, range checks are performed to assure an integer between -32767 and 32767.

Arguments

KFLD	(IN)	Alpha buffer containing user data input
LNGTH	(IN)	Length of KFLD, usually 11
VALUE	(OUT)	Real value returned to calling routine
INTG	(IN)	Integer flag, 0 = floating point, 1 = integer
NERR	(OUT)	Error flag, set to various values if an error occurs

SUBROUTINE ERROR

Abstract

Contains a set of error messages which appear in a reserved row on the CRT screen for illegal data entries.

Description

Displays any of three error messages in row 21. The '*ERROR*', and ' RE-ENTER' designations appear in columns 73-79 of the row of the data entry prompt and columns 62-70 of the row immediately below the input field. Selection of the particular error message displayed is controlled by GETVAL.

Arguments

ITYP	(IN)	Error message type to display
Y	(IN)	Row number of current data entry prompt
ERF	(OUT)	Set to 1 when error message is displayed. CLEAN will reset it to 0 after the messages are erased.

SUBROUTINE CLEAN

Abstract

Erases and cleans the screen of previous error messages.

Description

Routine is called by GETVAL when a user input error has been corrected. CLEAN clears the '*ERROR*', ' RE-ENTER', and any message residing in row number 21 of the CRT screen.

Arguments

Y	(IN)	Row number of current data entry prompt
ERF	(OUT)	Set to 0 upon exit to indicate messages have been cleared

SUBROUTINE BELL

Abstract

Produces an audible 'beep' to the terminal when a user input error is detected.

Description

Routine is called by ERROR. The ASCII bell code contained in IBELL is initialized by CRINIT.

Arguments

None.

SUBROUTINE CLRSCR

Abstract

Clears and erases all information on the CRT screen.

Description

Routine is called throughout program whenever clearing the screen to blanks is required. ICLEAR is set to the ASCII clear screen codes initialized in CRINIT.

SUBROUTINE INFO

Abstract

Displays failsafe information on last 3 rows of CRT screen.

Description

Row number 22 is always filled with underscores to create a line on the screen. Rows 23 and 24 contain operator instructions that may vary depending on the value of ITYP. Within the program system, calls to place various text in this area are performed whenever additional or revised operator instructions are required.

Arguments

ITYP	(IN)	Determines general operator instruction to appear. 1 = EXIT message only 2 = EXIT and TOP messages 3 = EXIT, TOP and BACK messages

SUBROUTINE ALLOW

Abstract

Displays various allowable value message formats following the data entry prompt.

Description

The allowable value message displayed under the user input data request prompt may take the following forms, depending on the setting on ITYP. ITYP (1-5) displays allowable values between a minimum and a maximum. The field width of the range is adjusted by the setting of ITYP. ITYP (6-7) displays the remaining allowable logic formats of 'greater than' and 'less than' a certain value.

Arguments

ITYP	(IN)	Selects allowable value format desired
Y	(IN)	Row number allowable value message placement
IR1	(IN)	Minimum allowable value (ALPHA)
IR2	(IN)	Maximum allowable value (ALPHA)

Chapter Bibliography

Hromadka II, T.V., Clements, J.M., and Guymon, G.L., "Guidelines for Interactive Software in Water Resources Engineering," Water Resources Bulletin, Feb.(1983c).

PROGRAM 24

```
C ---------------------------------------------------------------------
      SUBROUTINE CRINIT
C ---------------------------------------------------------------------
C
C     Routine to initialize the cursor control.
C     INITIALIZES FUNCTIONS OF BELL,CLEAR,CURSOR ADDRESSING
C     FOR LEAR SIEGLER TYPE TERMINALS (ADM3,3A,3A+,5,31);TELEVIDEO
C     AND SOROC SERIES
C
      COMMON /CURS/ IP(80),IESC,IBELL,ICLEAR(3)
C
      IBELL=7*256
C
      ICLEAR(1)=27*256
      ICLEAR(2)=43*256
      ICLEAR(3)=26*256
      IESC=27*256
C
      DO 101 I=1,80
      IP(I)=(I+31)*256
 101  CONTINUE
C
      RETURN
      END
```

PROGRAM 25

```
C ---------------------------------------------------------------------
      SUBROUTINE CURSOR(IX,IY)
C ---------------------------------------------------------------------
C
C     Routine to position the cursor anywhere on the screen
C
C     X is left to right, from 1 - 80
C     Y is up to down, from 1 - 24
C   FOR LEAR SIEGLER TERMINALS - LEAD IN SEQUENCE: ESC,=,Y,X
C
      COMMON/UNIT/WT,RT
      INTEGER WT,RT
      COMMON /CURS/ IP(80),IESC,IBELL,ICLEAR(3)
C
      WRITE (WT,10) IESC,IP(IY),IP(IX)
 10   FORMAT (A1,'=',2A1,Z)
C
      RETURN
      END
```

PROGRAM 26

```
C  -----------------------------------------------------------------
        SUBROUTINE GETVAL (IX,IY,INT,FMIN,FMAX,VAL,NCOND,MODE,NULL)
C  -----------------------------------------------------------------
C
C       GETS AND RETURNS A LEGAL VALUE FROM INPUT.
C       VALUE IS SYNTAX CHECKED, RANGE CHECKED, AND CHECKED FOR
C       OVERFLOW IN THE INTEGER CASE.
C
        INTEGER BUF(11),EXIT,TOP,BACK,WT,RT,ERF,BLK
        COMMON /UNIT/ WT,RT
        DATA EXIT/'E '/, TOP/'T '/,BACK/'B '/,BLK/'  '/,
     .  MA/'M '/
C
        NCOND=0
        ERF=0
10      CALL CURSOR(IX,IY)
        READ(RT,520) BUF
520     FORMAT(11A1)
C
        IF (BUF(1) .NE. EXIT) GO TO 20
        IF(MODE.LT.1) GO TO 55
        NCOND=1
        GO TO 999
C
20      IF(BUF(1) .NE. TOP) GO TO 30
        IF(MODE.LT.2) GO TO 55
        NCOND=2
        GO TO 999
C
30      IF(BUF(1) .NE. BACK) GO TO 70
        IF(MODE.LT.3 .OR. MODE.EQ.4) GO TO 55
        NCOND=3
        GO TO 999
C
70      IF(BUF(1).NE.MA)GO TO 40
        IF(MODE.LT.4)GO TO 55
        NCOND=4
        GO TO 999
C
40      CALL NUMCK(BUF,11,VAL,INT,NERR)
        IF(NERR.NE.3) GO TO 42
        IF(NULL.EQ.0) GO TO 55
        NCOND=5
        GO TO 999
42      CONTINUE
        IF(NERR .EQ.0) GO TO 50
        CALL ERROR(2,IY,ERF)
        GO TO 10
50      IF(VAL .GE. FMIN .AND. VAL .LE. FMAX) GO TO 60
        CALL ERROR(1,IY,ERF)
        GO TO 10
55      CALL ERROR(3,IY,ERF)
        GO TO 10
60      IF(ERF .EQ. 1) CALL CLEAN(IY,ERF)
C
999     CONTINUE
C
        RETURN
        END
```

PROGRAM 27

```
C  -----------------------------------------------------------------
       SUBROUTINE NUMCK (KFLD,LNGTH,VALUE,INTG,NERR)
C  -----------------------------------------------------------------
C
C      FREE FORMAT EXTRACTION FROM A1-ARRAY (KFLD)
C
C      KFLD      A1 ARRAY CONTAINING THE TARGET STRING
C      LNGTH     LENGTH OF KFLD ARRAY
C      VALUE     RETURNS A REAL VALUE DECODED FROM TARGET STRING
C      INTG      INTEGER FLAG     1=INTEGER
C      NERR      ERROR FLAG       1=NON NUMERIC CHARACTERS IN STRING
C                                 2=INTEGER OVERFLOW
C                                 3=BLANK ENTRY
C
       DIMENSION KFLD (LNGTH),IHOLD(15),IFMT(4)
       DATA IBK/'  '/
       NERR = 0
       VALUE = 0
       N1 = 0
       N2 = 0
       IF (LNGTH .LE.0) GO TO 1000
C
C  SCAN FIELD TO DELINEATE NON-BLANK CHARACTERS
C
       DO 130 I=1,LNGTH
       IF (KFLD(I) .NE. IBK) GO TO 120
       IF (N1 .EQ. 0) GO TO 130
       N2 = I-1
       GO TO 150
120    IF (N1.EQ.0) N1 = I
130    CONTINUE
       IF (N1.NE.0) GO TO 150
C  BLANK ENTRY
       VAL=0.
       NERR=3
       GO TO 1000
C
150    IF (N2.EQ.0) N2 = LNGTH
C
C  LOGIC FOR VALID REAL OR INTEGER CHARACTERS
C
       INUM = 0
       IBLK = 0
       ISGN = 0
       IPRD = 0
       NFLG = 0
       DO 200 I=N1,LNGTH
       IBY = I+(I-1)
       NCHR = BYTE(KFLD,IBY)
       IF(IBLK.EQ.1) GO TO 201
C      NUMBERS (CHECKED AGAINST OCTAL REPRESENTATION; "K")
       IF(NCHR .LT. 60K .OR. NCHR .GT. 71K) GO TO 202
               INUM=1
               ISGN=1
               GO TO 200
C      + OR - SIGN
202    IF(NCHR .NE. 53K .AND. NCHR .NE. 55K) GO TO 203
               IF(ISGN.EQ.1) GO TO 999
               ISGN=1
```

```
                 GO TO 200
C        PERIOD
203      IF(NCHR.NE.56K) GO TO 205
         IF(IPRD .EQ. 1) GO TO 999
                 ISGN=1
                 IPRD=1
                 GO TO 200
C        BLANK
205      IF (NCHR .NE. 40K) GO TO 999
                 IBLK=1
                 GO TO 200
201      IF(NCHR .NE. 40K) GO TO 999
200      CONTINUE
C        CHECKS FOR PERIOD OR SIGN IN LAST BYTE OF KFLD
         IF(IPRD.EQ.1 .AND. INUM.NE.1) GO TO 999
         IF(ISGN.EQ.1 .AND. INUM.NE.1) GO TO 999
C
C BUILD FORMAT
         ENCODE (IHOLD,600) (KFLD(I),I=N1,N2)
         N3=(N2-N1) +1
         ENCODE (IFMT,610) N3
         DECODE (IHOLD,IFMT) VALUE
C
C  INTEGER OVERFLOW CHECK
         IF(INTG .EQ. 0) GO TO 1000
         IF (VALUE.GE. -32767 .AND. VALUE.LE. 32767) GO TO 300
         NERR = 2
         VALUE = 0.
         GO TO 1000
C  CHECK FOR WHOLE NUMBER
300      IVALUE=VALUE
         REM=VALUE-IVALUE
         IF(REM.EQ.0.) GO TO 1000
         NERR=2
         VALUE=0.
         GO TO 1000
999      NERR = 1
C
C  FORMATS
C
600      FORMAT(30A1)
610      FORMAT(2H(F,I2,4H.0 ))
C
1000     RETURN
         END
```

PROGRAM 28

```
C  ----------------------------------------------------------------------
         SUBROUTINE ERROR (ITYP,Y,ERF)
C  ----------------------------------------------------------------------
C
C  DISPLAYS ERROR MESSAGES FOR ILLEGAL USER INPUT
C
         INTEGER Y,ERF
         INTEGER WT,RT
         COMMON /UNIT/WT,RT
         CALL BELL
C  CURSOR AT ERROR INDICATOR
         CALL CURSOR (73,Y)
         WRITE(WT,800)
C  CURSOR AT MESSAGE
         CALL CURSOR(1,21)
         IF(ITYP.EQ.1) WRITE(WT,801)
         IF(ITYP.EQ.2) WRITE(WT,804)
         IF(ITYP.EQ.3) WRITE(WT,805)
C
C CLEAR INPUT FIELD
         CALL CURSOR (62,Y)
         WRITE(WT,806)
         ERF=1
         CALL CURSOR(62,Y+1)
         WRITE(WT,807)
C
C  FORMATS
C
800      FORMAT('*ERROR*',Z)
801      FORMAT('*****VALUE ENTERED IS NOT WITHIN ACCEPTABLE RANGE. ',
     .   '              ',Z)
804      FORMAT('*****VALUE ENTERED CONTAINS NON-NUMERIC CHARACTERS.',
     .   '              ',Z)
805      FORMAT('*****I DO NOT UNDERSTAND.                            ',
     .   '              ',Z)
806      FORMAT(11(' '))
807      FORMAT('^RE-ENTER',Z)
C
         RETURN
         END
```

PROGRAM 29

```
C ----------------------------------------------------------------------
      SUBROUTINE CLEAN(Y,ERF)
C ----------------------------------------------------------------------
C
C  CLEANS THE ERROR MESSAGES AFTER USER ERROR IS CORRECTED
C
      INTEGER Y,ERF,UNIT,WT,RT
      COMMON /UNIT/WT,RT
C  CURSOR AT ERROR INDICATOR
      CALL CURSOR(73,Y)
      WRITE(WT,803)
C  CURSOR AT MESSAGE
      CALL CURSOR(1,21)
      WRITE(WT,804)
      ERF = 0
      CALL CURSOR(62,Y+1)
      WRITE(WT,805)
C
C  FORMATS
C
803   FORMAT(7(' '),Z)
804   FORMAT(56(' '),Z)
805   FORMAT(9(' '),Z)
C
      RETURN
      END
```

PROGRAM 30

```
C ----------------------------------------------------------------------
      SUBROUTINE BELL
C ----------------------------------------------------------------------
C
C  GENERAL BELL FUNCTION; INITIALIZED BY CRINIT
C
      INTEGER WT,RT
C
      COMMON/CURS/IP(80),IESC,IBELL,ICLEAR(3)
      COMMON/UNIT/WT,RT
C
      WRITE(WT,600)IBELL
600   FORMAT(A1,Z)
C
      RETURN
      END
```

PROGRAM 31

```
C  ------------------------------------------------------------------
       SUBROUTINE CLRSCR
C  ------------------------------------------------------------------
C
C  GENERAL CLEAR FUNCTION; INITIALIZED BY CRINIT
C
       INTEGER WT,RT
C
       COMMON/CURS/IP(80),IESC,IBELL,ICLEAR(3)
       COMMON/UNIT/WT,RT
C
       WRITE(WT,600)ICLEAR
600    FORMAT(3A1,Z)
C
       RETURN
       END
```

PROGRAM 32

```
C  ------------------------------------------------------------------
       SUBROUTINE INFO(ITYP)
C  ------------------------------------------------------------------
C
C  DISPLAYS STANDARD OPERATOR INFORMATION ON LINES 22-24
C
       COMMON /UNIT/ WT,RT
       INTEGER WT,RT
C
       CALL CURSOR(1,22)
       WRITE(WT,600)
       CALL CURSOR(5,23)
       WRITE(WT,601)
       IF(ITYP.EQ.1) GO TO 999
       CALL CURSOR(33,23)
       WRITE(WT,602)
       IF(ITYP.EQ.2) GO TO 999
       CALL CURSOR(33,24)
       WRITE(WT,603)
999    CONTINUE
C
C  FORMATS
C
600    FORMAT('____________________________________________________',
     .  '________________________________',Z)
601    FORMAT('TYPE: EXIT to leave program',Z)
602    FORMAT('; TOP  to go to top of page         ',Z)
603    FORMAT('; BACK to go back one page         ',Z)
C
       RETURN
       END
```

PROGRAM 33

```
C -------------------------------------------------------------------
      SUBROUTINE ALLOW (ITYP,Y,IR1,IR2)
C -------------------------------------------------------------------
C
C  DISPLAYS ALLOWABLE VALUE MESSAGES
C
      DIMENSION IR1(5),IR2(5)
      INTEGER WT,RT
      COMMON /UNIT/WT,RT
C  CURSOR AT PLACEMENT OF ALLOWABLE MESSAGE
      CALL CURSOR(5,Y)
C
      IF(ITYP.EQ.1) WRITE(WT,801) IR1(1),(IR2(I),I=1,3)
      IF(ITYP.EQ.2) WRITE(WT,802) IR1(1),(IR2(I),I=1,5)
      IF(ITYP.EQ.3) WRITE(WT,803)(IR1(I),I=1,3),(IR2(I),I=1,3)
      IF(ITYP.EQ.4) WRITE(WT,804)(IR1(I),I=1,3),(IR2(I),I=1,5)
      IF(ITYP.EQ.5) WRITE(WT,805)(IR1(I),I=1,5),(IR2(I),I=1,5)
      IF(ITYP.EQ.6) WRITE(WT,806)(IR1(I),I=1,3)
      IF(ITYP.EQ.7) WRITE(WT,807)(IR1(I),I=1,3)
C
C  FORMATS
C
801   FORMAT(':ALLOWABLE VALUES ARE [',A1,'] TO [',3A2,']')
802   FORMAT(':ALLOWABLE VALUES ARE [',A1,'] TO [',5A2,']')
803   FORMAT(':ALLOWABLE VALUES ARE [',3A2,'] TO [',3A2,']')
804   FORMAT(':ALLOWABLE VALUES ARE [',3A2,'] TO [',5A2,']')
805   FORMAT(':ALLOWABLE VALUES ARE [',5A2,'] TO [',5A2,']')
806   FORMAT(':ALLOWABLE VALUES ARE GREATER THAN [',3A2,']')
807   FORMAT(':ALLOWABLE VALUES ARE LESS THAN [',3A2,']')
C
      RETURN
      END
```

Appendix A - Lag Relationships

		DRAINAGE AREA SQ. MI.	L MILES	L_{CA} MILES	S FT./MI.	LAG HOURS
1.	SAN GABRIEL RIVER AT SAN GABRIEL DAM NO. 1*	162	23.2	11.6	350	3.3
2.	WEST FORK SAN GABRIEL RIVER AT SAN GABRIEL DAM NO. 2	40.4	9.3	4.2	450	1.6
3.	SANTA ANITA CREEK AT SANTA ANITA DAM	10.8	5.8	2.5	690	1.1
4.	SAN DIMAS CREEK AT SAN DIMAS DAM	16.2	8.6	4.8	440	1.5
5.	EATON WASH AT EATON DAM	9.5	7.3	4.4	600	1.3
6.	MURRIETA CREEK AT TEMECULA	220	27.2	10.3	95	4.0
7.	SANTA CLARA RIVER NEAR SAUGUS	355	36.0	15.8	140	5.6
8.	TEMECULA CREEK AT PAUBA CANYON**	168	26.0	11.3	150	3.7
9.	SANTA MARGARITA RIVER NEAR FALLBROOK	645	46.0	22.0	105	7.3
10.	EAST FULLERTON CREEK AT FULLERTON DAM	3.1	3.2	1.7	140	0.8
11.	TUJUNGA CREEK AT BIG TUJUNGA DAM NO. 1	81.4	15.1	7.3	290	2.5
12.	SANTA MARGARITA RIVER AT YSIDORA	740	61.2	34.3	85	9.5

* EXCLUDES AREA ABOVE SAN GABRIEL DAM NO. 2

** PALOMAR MOUNTAIN PORTION. ENTIRE AREA IS 319 SQUARE MILES, OF WHICH 151 SQUARE MILES DID NOT CONTRIBUTE APPRECIABLE FLOOD FLOWS DURING THE 1937 AND 1940 OCCURRENCES.

TERMINOLOGY

L = LENGTH OF LONGEST WATERCOURSE.

L_{CA} = LENGTH OF LONGEST WATERCOURSE, MEASURED UPSTREAM, TO POINT OPPOSITE CENTER OF AREA.

S = OVER-ALL SLOPE OF DRAINAGE AREA BETWEEN HEADWATERS AND COLLECTION POINT.

LAG = ELAPSED TIME FROM BEGINNING OF UNIT RAINFALL TO INSTANT THAT SUMMATION HYDROGRAPH REACHES 50% OF ULTIMATE DISCHARGE.

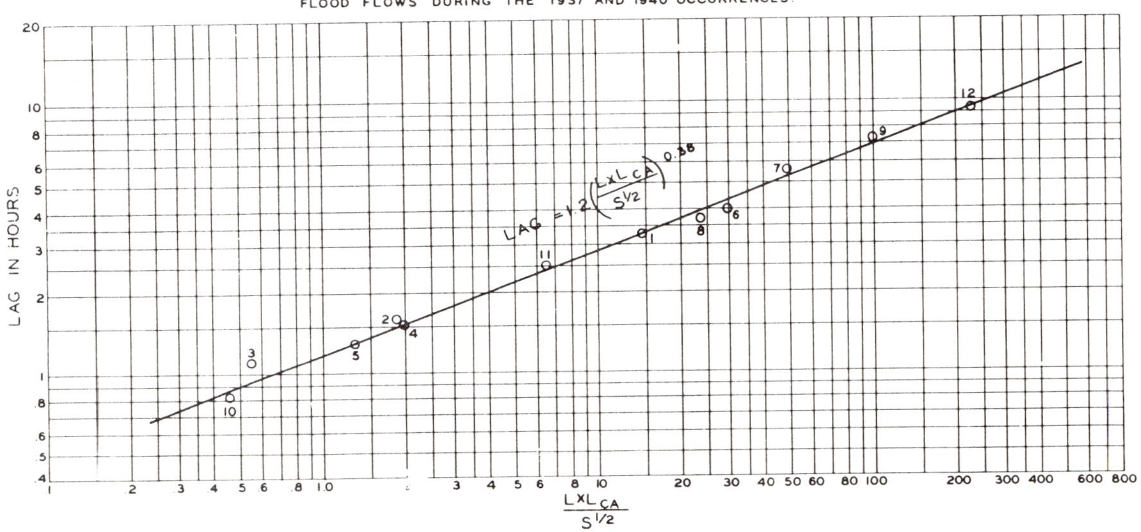

LAG RELATIONSHIP FOR WATERSHEDS SOUTHERN CALIFORNIA

APPENDIX B-HARDWARE EQUIPMENT LIST

A typical hardware configuration required for the provided program systems is essentially a standard configuration for an Apple II[1] series computer, with two disk drives, monitor, printer, and a Microsoft Softcard[1] installed. The following lists the recommended peripheral interface boards that are compatible with the supplied software listings:

1. Apple II series computer.

2. Disk drives (2), Apple brand recommended.

3. NEC 1201 high resolution green screen monitor.

4. 80 column card - Videx Videoterm or Smarterm for Apple II plus, Apple 80 column card for Apple IIe.

5. Microsoft Softcard CP/M system.

6. Printer and interface card - OKIDATA 93 160cps wide carriage printer with a Microbuffer or Wizzard internal 16K buffer card interface for simultaneous printing and processing.

7. Microsoft RAMCard or Apple Language card for additional 16K memory (for Apple II plus only). Apple IIe has the additional memory required.

[1] Apple II is a trademark of Apple Computer, Inc. CP/M is a trademark of Digital Research, Inc. Softcard is a trademark of Microsoft.